ENERGY SCIENCE, ENGINEERING AND TECHNOLOGY

NATURAL GAS: OUTLOOKS AND OPPORTUNITIES

ENERGY SCIENCE, ENGINEERING AND TECHNOLOGY

Additional books in this series can be found on Nova's website under the Series tab.

Additional E-books in this series can be found on Nova's website under the E-books tab.

Energy Science, Engineering and Technology

Natural Gas: Outlooks and Opportunities

Lucas N. Montauban
Editor

Nova Science Publishers, Inc.
New York

Copyright © 2011 by Nova Science Publishers, Inc.

All rights reserved. No part of this book may be reproduced, stored in a retrieval system or transmitted in any form or by any means: electronic, electrostatic, magnetic, tape, mechanical photocopying, recording or otherwise without the written permission of the Publisher.

For permission to use material from this book please contact us:
Telephone 631-231-7269; Fax 631-231-8175
Web Site: http://www.novapublishers.com

NOTICE TO THE READER

The Publisher has taken reasonable care in the preparation of this book, but makes no expressed or implied warranty of any kind and assumes no responsibility for any errors or omissions. No liability is assumed for incidental or consequential damages in connection with or arising out of information contained in this book. The Publisher shall not be liable for any special, consequential, or exemplary damages resulting, in whole or in part, from the readers' use of, or reliance upon, this material. Any parts of this book based on government reports are so indicated and copyright is claimed for those parts to the extent applicable to compilations of such works.

Independent verification should be sought for any data, advice or recommendations contained in this book. In addition, no responsibility is assumed by the publisher for any injury and/or damage to persons or property arising from any methods, products, instructions, ideas or otherwise contained in this publication.

This publication is designed to provide accurate and authoritative information with regard to the subject matter covered herein. It is sold with the clear understanding that the Publisher is not engaged in rendering legal or any other professional services. If legal or any other expert assistance is required, the services of a competent person should be sought. FROM A DECLARATION OF PARTICIPANTS JOINTLY ADOPTED BY A COMMITTEE OF THE AMERICAN BAR ASSOCIATION AND A COMMITTEE OF PUBLISHERS.

Additional color graphics may be available in the e-book version of this book.

Library of Congress Cataloging-in-Publication Data

Natural gas : outlooks and opportunities / editor, Lucas N. Montauban.
    p. cm.
  Includes bibliographical references and index.
  ISBN 978-1-61324-132-5 (hardcover : alk. paper) 1. Gas as fuel. 2. Gas distribution. I. Montauban, Lucas N.
  TP350.N36455 2011
  333.8'233--dc22
                          2011009070

Published by Nova Science Publishers, Inc. † New York

CONTENTS

Preface		vii
Chapter 1	Global Natural Gas: A Growing Resource *Michael Ratner*	1
Chapter 2	Displacing Coal with Generation from Existing Natural Gas-Fired Power Plants *Stan Mark Kaplan*	17
Chapter 3	Federal Oil and Gas Leases: Opportunities Exist to Capture Vented and Flared Natural Gas, which Would Increase Royalty Payments and Reduce Greenhouse Gases *United States Government Accountability Office Report to Congressional Requesters*	47
Chapter 4	The Alaska Natural Gas Pipeline: Background, Status, and Issues for Congress *Paul W. Parfomak*	85
Chapter 5	Liquefied Natural Gas (LNG) Import Terminals: Siting, Safety, and Regulation *Paul W. Parfomak and Adam Vann*	107
Chapter 6	Natural Gas Passenger Vehicles: Availability, Cost, and Performance *Brent D. Yacobucci*	137
Chapter Sources		143
Index		145

PREFACE

Natural gas is considered a potential bridge fuel to a low carbon economy because it is cleaner burning than its hydrocarbon rivals coal and oil. Natural gas combustion emits about two-thirds less carbon dioxide than coal and one-quarter less than oil when consumed in a typical electric power plant. Additionally, improved methods to extract natural gas from certain shale formations has significantly increased the resource profile of the United States, which has spurred other countries to try to develop shale gas. If the United States and other countries can bring large new volumes of natural gas to market, then natural gas could play a larger role in the world's economy. This book examines key aspects of global natural gas markets, including supply and demand, as well as major U.S. developments.

Chapter 1- The role of natural gas in the U.S. economy is expected to be a major part of the debate over energy policy in the 112th Congress. This report briefly explains key aspects of global natural gas markets, including supply and demand, as well as major U.S. developments.[1]

Natural gas is considered a potential bridge fuel to a low carbon economy because it is cleaner burning than its hydrocarbon rivals coal and oil. Natural gas combustion emits about two-thirds less carbon dioxide than coal and one-quarter less than oil when consumed in a typical electric power plant.[2] Natural gas combustion also emits less particulate matter, sulfur dioxide, and nitrogen oxides than coal or oil. Additionally, improved methods to extract natural gas from certain shale formations has significantly increased the resource profile of the United States, which has spurred other countries to try to develop shale gas. If the United States and other countries can bring large new volumes of natural gas to market, then natural gas could play a larger role in the world's economy. Several key factors will determine whether significant new quantities of natural gas come to market, particularly unconventional natural gas resources.[3] These factors include price, technical capability, environmental concerns, and political considerations. Many countries, both producing and consuming, are watching how the development of U.S. unconventional natural gas resources evolves.

Chapter 2- Reducing carbon dioxide emissions from coal plants is a focus of many proposals for cutting greenhouse gas emissions. One option is to replace some coal power with natural gas generation, a relatively low carbon source of electricity, by increasing the power output from currently underutilized natural gas plants.

This report provides an overview of the issues involved in displacing coal-fired generation with electricity from existing natural gas plants. This is a complex subject and the report does not seek to provide definitive answers. The report aims to highlight the key issues that Congress may want to consider in deciding whether to rely on, and encourage, displacement of coal-fired electricity with power from existing natural gas plants.

Chapter 3- The Department of the Interior (Interior) leases public lands for oil and natural gas development, which generated about $9 billion in royalties in 2009. Some gas produced on these leases cannot be easily captured and is released (vented) directly to the atmosphere or is burned (flared). This vented and flared gas represents potential lost royalties for Interior and contributes to greenhouse gas emissions.

GAO was asked to (1) examine available estimates of the vented and flared natural gas on federal leases, (2) estimate the potential to capture additional gas with available technologies and associated potential increases in royalty payments and decreases in greenhouse gas emissions, and (3) assess the federal role in reducing venting and flaring. In addressing these objectives, GAO analyzed data from Interior, the Environmental Protection Agency (EPA), and others and interviewed agency and industry officials.

Chapter 4- Constructing a natural gas pipeline from Alaska's North Slope to the lower-48 states has been a government priority—periodically—for more than four decades. Beginning with the Alaska Natural Gas Transportation Act of 1976, Congress has repeatedly affirmed a national need for an Alaska gas pipeline. In remarks to the press President Obama has likewise described an Alaska gas pipeline as "a project of great potential ... as part of a comprehensive energy strategy."

Chapter 5- Liquefied natural gas (LNG) is a hazardous fuel shipped in large tankers to U.S. ports from overseas. While LNG has historically made up a small part of U.S. natural gas supplies, rising price volatility, and the possibility of domestic shortages have significantly increased LNG demand. To meet this demand, energy companies have proposed new LNG import terminals throughout the coastal United States. Many of these terminals would be built onshore near populated areas.

Chapter 6- Higher gasoline prices in recent years and concerns over U.S. oil dependence have raised interest in natural gas vehicles (NGVs). Use of NGVs for personal transportation has focused on compressed natural gas (CNG) as an alternative to gasoline. Consumer interest has grown, both for new NGVs as well as for conversions of existing personal vehicles to run on CNG. This report finds that the market for natural gas passenger vehicles will likely remain limited unless the differential between natural gas and gasoline prices remains high in order to offset the higher purchase price for an NGV. Conversions of existing vehicles will also continue to be restricted unless the Clean Air Act (CAA) is amended or if the Environmental Protection Agency (EPA) makes changes to its enforcement of the CAA.

Natural Gas: Outlooks and Opportunities
Editor: Lucas N. Montauban

ISBN: 978-1-61324-132-5
© 2011 Nova Science Publishers, Inc.

Chapter 1

GLOBAL NATURAL GAS: A GROWING RESOURCE[*]

Michael Ratner

INTRODUCTION

The role of natural gas in the U.S. economy is expected to be a major part of the debate over energy policy in the 112th Congress. This report briefly explains key aspects of global natural gas markets, including supply and demand, as well as major U.S. developments.[1]

Natural gas is considered a potential bridge fuel to a low carbon economy because it is cleaner burning than its hydrocarbon rivals coal and oil. Natural gas combustion emits about two-thirds less carbon dioxide than coal and one-quarter less than oil when consumed in a typical electric power plant.[2] Natural gas combustion also emits less particulate matter, sulfur dioxide, and nitrogen oxides than coal or oil. Additionally, improved methods to extract natural gas from certain shale formations has significantly increased the resource profile of the United States, which has spurred other countries to try to develop shale gas. If the United States and other countries can bring large new volumes of natural gas to market, then natural gas could play a larger role in the world's economy. Several key factors will determine whether significant new quantities of natural gas come to market, particularly unconventional natural gas resources.[3] These factors include price, technical capability, environmental concerns, and political considerations. Many countries, both producing and consuming, are watching how the development of U.S. unconventional natural gas resources evolves.

Key Points:

- Natural gas is likely to play a greater role in the world energy mix given its growing resource[4] base and its relatively low carbon emissions compared to other fossil fuels.
- The world used over 100,000 billion cubic feet (bcf) of natural gas in 2009, of which the United States consumed almost 23,000 bcf, the most of any country. Between

---

[*] This is an edited, reformatted and augmented version of a Congressional Research Services publication, dated December 22, 2010.

2008 and 2009, world consumption declined about 2.6%, while U.S. consumption dropped 1.6%, or 388 bcf.

- In 2009, almost 84% of the natural gas the United States consumed was from domestic production. Another 14 % of consumption was met with Canadian imports. Liquefied natural gas[5] (LNG), mainly from Trinidad & Tobago and Egypt, comprised just 2% of consumption.
- U.S. unconventional natural gas reserves[6] and production, particularly shale gas, have grown rapidly in recent years. In 2009, shale gas reserves increased 76%, while production rose 47%.[7] The new shale gas resources have changed the U.S. natural gas position from net importer to potentially a net exporter. Other countries are now exploring their own shale gas resources.

NATURAL GAS CONSUMPTION

Key Global Consumers

In 2009, the world consumed almost 104,000 bcf of natural gas—24% of total global energy consumption and 27% of U.S. needs. The United States was the world's largest consumer of natural gas, accounting for 22,849 bcf, or 22%, of global consumption (Figure 1). Consumption of natural gas declined both globally and in the United States by about 2% last year—the most rapid decline on a global basis on record[8]—which can be mainly attributed to the economic downturn.

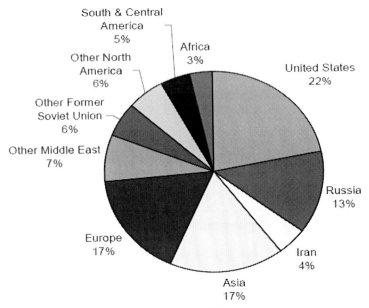

Source: BP Statistical Review of World Energy 2010, p. 27.

Figure 1. Global Natural Gas Consumption; Total global consumption was 103,825 bcf in 2009.

Electric power generation, residential and commercial uses, and industrial uses each account for about one-third of U.S. natural gas consumption. In 2009, electric power generation, which was the only sector to increase its gas usage above 2008 levels, led the consuming sectors with 6,900 bcf of natural gas, or a 3% rise.[9] Russia and Iran, the second and third largest consumers of natural gas, both subsidize natural gas usage, which increases their consumption. China, the fifth largest consumer last year, with a global share of just over 3%, is viewed as a growing market for natural gas, which currently comprises less than 4% of China's primary energy use.

NATURAL GAS SUPPLY AND TRADE

Natural Gas Reserves Growing

Global proved natural gas reserves[10]—natural gas that has been discovered and can be expected to be economically produced—amounted to 6,621,153 bcf, which correlates to over a 60-year supply at current production levels. New reserves are developed every year as existing reserves are consumed, so that the ratio between the world's reserves and global production has remained around 60 years since 1980. Natural gas reserves have grown about 6% since 2007, demonstrating the success of exploration and improved recovery techniques.[11]

In 2009, U.S. natural gas reserves were 244,731 bcf (the value used for international comparisons in this report),[12] or about 12 years' supply at current production levels. However, a recently released report by the U.S. EIA revised U.S. reserves upward to 284,000 bcf, primarily driven by shale gas additions.[13] The improvements in development of shale gas resources over the last two years have changed the U.S. supply profile for natural gas. In June 2009, the Potential Gas Committee[14] released its biennial assessment of U.S. natural gas resources, including reserves, which total over 1,765,735 bcf according to the report, an increase of almost 40% over the last assessment. The increase is attributed to a re-evaluation of shale gas resources, primarily in the Appalachian basin, Mid-Continent (including parts of Arkansas, Oklahoma, and Texas), Gulf Coast, and Rocky Mountain areas.

Shale gas accounted for 21% of U.S. natural gas reserves in 2009, up from 14% in 2008.[15] Nevertheless, questions still remain about the size of U.S. shale gas resources (which by most estimates is more than current U.S. natural gas reserves); the price level required to sustain their development; and whether there are technical, environmental, or political factors that might limit their development. The use and disposition of water in an industry process called hydraulic fracturing[16] is the main issue facing companies and regulators. The EPA is undertaking a study to determine any adverse effects of the practice on water supplies or other environmental areas.[17]

Globally, over half of the world's proved natural gas reserves are controlled by the top-10 government owned companies, with all but one being 100% state owned. Russia's Gazprom is majority owned by the state and acts as an arm of the government. Iran's National Iranian Oil Company is the largest reserve holder.

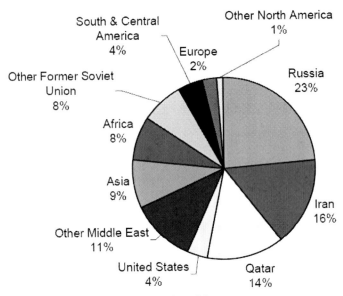

Source: BP Statistical Review of World Energy 2010, p. 22.

Figure 2. Global Natural Gas Reserves; Global natural gas reserves were 6,621,153 bcf in 2009.

Global Natural Gas Market Becoming More Integrated

Although some natural gas is traded around the world, most natural gas is predominantly consumed in the country where it is produced (Figure 4). Only about 30% of natural gas is traded internationally, mostly within regional markets. Nevertheless, the amount of natural gas traded has been increasing over the last five years. Natural gas is transported in two ways: by pipeline and as a liquid in tankers, which is an expensive process. Liquefaction capacity has increased 30% since 2008, and trade in LNG has grown almost 30% since 2005. International pipeline trade is up almost 20% since 2005. Pipelines transport gas between two fixed points, while LNG provides flexibility in the final destination.

Almost all natural gas that is traded internationally is under long-term contracts, usually 20 years, whether it is by pipeline or as LNG. This is primarily because natural gas transportation is expensive and the long-term contracts are needed to finance construction of the transport facilities. Sometimes LNG consumers do not require the entire amount of natural gas in their contracts, and LNG producers can then sell that natural gas to other consumers on a one-time or short-term basis (e.g., sell it on "spot").

Russia is the world's largest natural gas exporter, primarily through its massive pipeline network to Europe. Russia opened its first LNG export terminal in 2009, primarily targeted at the Asian market, to give it flexibility in its exports. Qatar is the leading exporter of LNG, accounting for 20% of world LNG trade, with exports going to 15 countries. Europe is the largest importing region of natural gas, receiving most of its imports by pipeline from Russia, Norway, and Algeria. Asia, the most import-dependent region, relies mostly on LNG, although China is actively pursuing pipeline projects with certain neighbors and opened its first import pipeline from Turkmenistan via Uzbekistan and Kazakhstan at the end of last year.

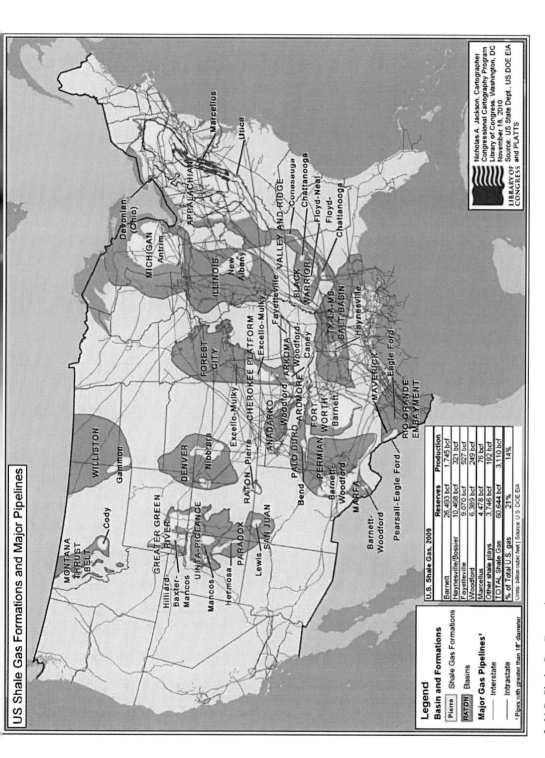

Figure 3. U.S. Shale Gas Formations.

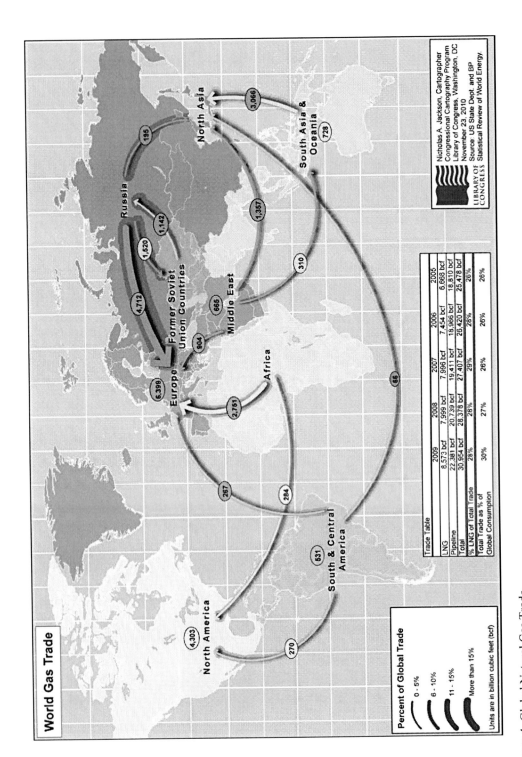

Figure 4. Global Natural Gas Trade.

Natural Gas Exporting Countries Forum Still Ineffective

The Gas Exporting Countries Forum (GECF), also referred to as gas OPEC, is a nascent cartel organization based in Qatar comprising 11 natural gas producing countries (**Table 1**). The GECF was formed in 2001, but only signed an organizing charter in December 2008. It controls 34% of global natural gas production and 44% of natural gas traded. Given the U.S. resource base of natural gas, it is highly unlikely that the GECF could significantly affect U.S. natural gas consumption within the next five years or, most likely, longer. Canada, by far the largest U.S. source of imported natural gas, is not a member of the GECF. Europe is probably most vulnerable to the cartel, as more than half its imports come from cartel members, particularly Russia and Algeria. Nevertheless, the current structure of natural gas markets (i.e., long-term contracts and pipelines connecting individual sellers to specific buyers) is not conducive to supply or price manipulation, and significant changes would need to be made to how natural gas is bought and sold before the GECF could have influence.

Table 1. GECF Natural Gas Statistics 2009
units = bcf

	Reserves	Production	LNG	Pipeline	Total
Algeria*	158,916	2,875	738	1,122	1,860
Bolivia	25,073	434	0	346	346
Egypt	77,339	2,214	453	194	647
Equatorial Guinea	4,238	221	167	0	167
Iran*	1,045,668	4,633	0	200	200
Libya*	54,385	540	25	324	349
Nigeria*	185,402	879	0	565	565
Qatar*	895,934	3,154	1,746	662	2,408
Russia	1,567,266	18,629	233	6,232	6,466
Trinidad & Tobago	15,538	1,434	697	0	697
Venezuela*	200,234	985	0	0	0
TOTAL GECF	4,229,995	35,999	4,059	9,646	13,705
% of World	64%	34%	47%	43%	44%

Sources: BP Statistical Review of World Energy 2010 and Cedigaz statistical databases.
Note: * denotes a member of OPEC.

Production Widespread

Overall, global natural gas production decreased 2.1% last year, the first decline on record. This was primarily driven by the reduction in demand resulting from the widespread economic downturn.

The United States surpassed Russia as the world's largest natural gas producer last year for the first time since 2001. The success of the United States to date and the potential for further shale gas development has initiated an evaluation by most countries of their possible natural gas resources. However, outside of Canada, whose shale gas industry is developing alongside that of the United States, it is unlikely that commercial production will be achieved before the end of the decade. Most countries looking at shale gas do not have the data, technology, or equipment required to evaluate their shale gas resources, let alone successfully exploit it, at this point.

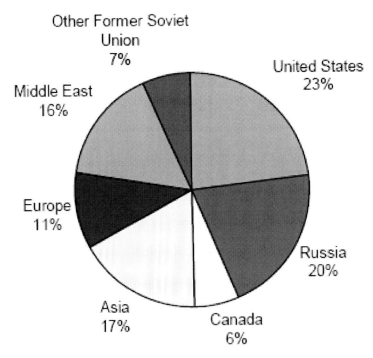

Source: BP Statistical Review of World Energy 2010, p. 24.

Figure 5. Global Natural Gas Production; Global natural gas production was 105,485 bcf in 2009.

Natural Gas Prices Remain Low

The market price for natural gas has been relatively low compared to the contract price in more competitive markets. The price of natural gas in the United States, Canada, and the United Kingdom is set by the market, with centers or hubs providing buyers and sellers with competitive price data. The most well-known hub in the United States is the Henry Hub in Erath, LA, which is where multiple interstate and intrastate natural gas pipelines interconnect. In the United States, there are various prices for natural gas depending upon the consumer. Residential[18] consumers pay the highest price, followed by commercial users.

Outside the United States, Canada, and the United Kingdom, almost all wholesale natural gas is sold under long-term contracts. The price of natural gas within the contracts is commonly determined by a formula that links the natural gas price to the price of crude oil or some oil-based product. Although in many markets natural gas no longer competes against oil-based products, this vestige of the contracts has not disappeared. Over the last several years, the disparity between contract prices and spot prices has raised the pressure on producers to do away with this concept (**Figure 6**). Producers have been reluctant, as oil prices are much higher than natural gas prices and the contract prices have been propped up by the difference. Nevertheless, some producers have started incorporating the spot price for natural gas into their pricing formulas. The price differences in **Figure 6** reflect the regional nature of the natural gas industry and the disparity between contract and spot prices. Asia, in particular, has been willing to pay high prices to secure its natural gas supplies.

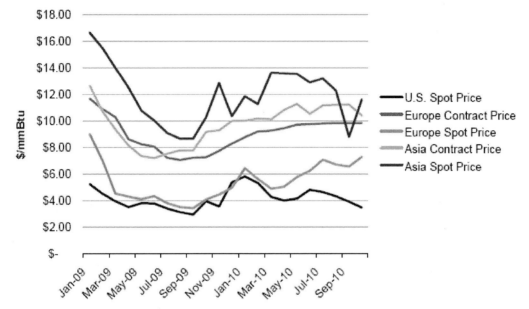

Source: PIRA Energy, November 8, 2010.
Notes: Contract price is a long-term price between a buyer and a seller, while the spot price is a short-term market price.

Figure 6. Global Natural Gas Prices; Units = U.S. dollars per million British thermal unit.

There are two other contract concepts that are worth highlighting: take-or-pay clauses and destination clauses. The take-or-pay clause does exactly what it says. A buyer of natural gas must pay the seller regardless of whether it actually receives the natural gas. Typically in contracts, buyers must purchase at least 80% of the total volume of natural gas contracted. For example, if a contract is for 100 bcf, but the buyer only needs 80 bcf, then that is all it pays for, but if the buyer only needs 50 bcf, it still must pay for an additional 30 bcf even if it cannot use it. A destination clause allows a cargo to be redirected to a different buyer. This clause was not common until recent years and contributes to a more efficient market.

MAJOR STATUTES

Table 2. Existing Legislation Governing Natural Gas in the United States

Act	Citation	Purpose
Natural Gas Act	15B U.S.C. § 717 et seq.	Governs siting of interstate natural gas pipelines and interstate transmission of natural gas. Also gives authority to DOE for imports and exports of liquefied natural gas (LNG).
Natural Gas Wellhead Decontrol Act of 1989	15 U.S.C. § 3301 et seq.	Removed remaining price ceilings on natural gas sales.
Mineral Leasing Act of 1920	30 U.S.C. § 181 et seq.	Governs leasing activity on federal lands, including leases for purposes of oil and natural gas exploration and production.
Outer Continental Shelf Lands Act	43 U.S.C. § 1331 et seq.	Governs activities on Outer Continental Shelf, including leasing for purposes of oil and natural gas exploration and production.
Natural Gas Pipeline Safety Act of 1968	P.L. 90-481a	Authorizes DOT to regulate pipeline transportation of natural gas and other gases as well as the transportation and storage of LNG.
Natural Gas Policy Act of 1978	P.L. 95-621	Gave FERC authority over intrastate and interstate natural gas production. The act also set price ceilings for natural gas.
Hazardous Liquid Pipeline Safety Act of 1979	P.L. 96-129a	Authorizes DOT to regulate pipeline transportation of hazardous liquids, including crude oil, petroleum products, anhydrous ammonia, and carbon dioxide.
Homeland Security Act of 2002	P.L. 107-296	Incorporated the Transportation Security Administration (TSA), which has jurisdiction for natural gas pipeline security, into the Department of Homeland Security.
Pipeline Safety Improvement Act of 2002	P.L. 107-355a	Strengthens federal pipeline safety programs, state oversight of pipeline operators, and public education regarding pipeline safety.

Source: Compiled by the Congressional Research Service (CRS).
Notes: The above list is not exhaustive, but highlights important statutes that relate to natural gas.
[a.] The Natural Gas Pipeline Safety Act of 1968, the Hazardous Liquid Pipeline Safety Act of 1979, and the Pipeline Safety Improvement Act of 2002 are re-codified at 40 U.S.C. Ch. 601.

LOOKING FORWARD

Is it finally time for natural gas to take center stage as a primary energy source? That is the main question confronting the natural gas industry over the next decade. The International Energy Agency (IEA) projects natural gas use to grow in all three of its scenarios out to 2035 in its recently released World Energy Outlook 2010.[19] Most of the new demand for natural gas is projected to come from non-OECD countries, primarily China and the Middle East. The electric power sector leads the growth in natural gas demand due to several factors, including relatively low prices, lower capital costs, and competitive financing of projects. Government policies, particularly in regard to carbon dioxide emissions, will be a key factor in determining the rate of growth of natural gas usage.

Natural gas production will increase to meet the rise in demand with growth projected in every region except Europe. Unconventional gas resources—coalbed methane, shale gas, and tight gas—will comprise 19% of production by 2035 according to the IEA report. Correspondingly, trade of natural gas is also forecast to expand, with Chinese imports growing the most.

Natural gas is likely to be addressed in multiple areas by the 112th Congress. It is one of the fuels included in a clean energy standard, particularly as a replacement for coal-fired electric power generation. Natural gas also factors into discussions on climate change, as it is the lowest carbon emitting fossil fuel per unit of energy produced when burned. Production of natural gas is included in any legislation related to drilling activity in the United States. Possible new regulations by the U.S. Environmental Protection Agency (EPA) will also impact the natural gas industry, especially shale gas development.

APPENDIX A. GLOBAL NATURAL GAS CONSUMPTION (2009)

Rank	Country	Consumption (bcf)	Share of World
1	United States	22,834	22%
2	Russia	13,762	13%
3	Iran	4,651	4%
4	Canada	3,344	3%
5	China	3,221	3%
6	Japan	3,087	3%
7	United Kingdom	3,055	3%
8	Germany	2,755	3%
9	Saudi Arabia	2,737	3%
10	Italy	2,529	2%
11	Mexico	2,458	2%
12	UAE	2,087	2%
13	India	1,833	2%
14	Uzbekistan	1,720	2%
15	Ukraine	1,660	2%
16	Argentina	1,522	1%
17	France	1,504	1%
18	Egypt	1,501	1%
19	Thailand	1,384	1%
20	Netherlands	1,374	1%
	Rest of World	24,823	24%
	Global Total	103,839	100%

Source: BP Statistical Review of World Energy 2010, p. 27.

APPENDIX B. GLOBAL NATURAL GAS RESERVES (2009)

Rank	Country	Reserves (bcf)	Share of World
1	*Russia*	1,567,266	24%
2	*Iran*	1,045,668	16%
3	*Qatar*	895,934	14%
4	Turkmenistan	286,049	4%
5	Saudi Arabia	279,692	4%
6	United States	244,731	4%
7	UAE	227,074	3%
8	*Venezuela*	200,234	3%
9	*Nigeria*	185,402	3%
10	*Algeria*	158,916	2%
11	Indonesia	112,301	2%
12	Iraq	111,948	2%
13	Australia	108,769	2%
14	China	86,874	1%
15	Malaysia	84,049	1%
16	*Egypt*	77,339	1%
17	Norway	72,395	1%
18	Kazakhstan	64,273	1%
19	Kuwait	62,860	1%
20	Canada	61,801	1%
	Rest of World	692,521	10%
	Global Total	6,621,153	100%

Source: BP Statistical Review of World Energy 2010, p. 22.
Note: GECF Member.

APPENDIX C. GLOBAL NATURAL GAS PRODUCTION (2009)

Rank	Country	Production (bcf)	Share of World
1	United States	20,956	20%
2	*Russia*	18,629	18%
3	Canada	5,700	5%
4	*Iran*	4,633	4%
5	Norway	3,655	3%
6	*Qatar*	3,154	3%
7	China	3,009	3%
8	*Algeria*	2,875	3%
9	Saudi Arabia	2,737	3%
10	Indonesia	2,539	2%
11	Uzbekistan	2,274	2%
12	*Egypt*	2,214	2%
13	Malaysia	2,214	2%
14	Netherlands	2,214	2%
15	United Kingdom	2,105	2%
16	Mexico	2,055	2%
17	UAE	1,723	2%
18	Australia	1,494	1%
19	Argentina	1,462	1%
20	*Trinidad & Tobago*	1,434	1%
	Rest of World	18,410	17%
	Global Total	105,485	100%

Source: BP Statistical Review of World Energy 2010, p. 24.
Notes: GECF Member.

APPENDIX D. U.S. NATURAL GAS IMPORTS AND EXPORTS

Table D-1. U.S. Imports of Natural Gas units = billion cubic feet (bcf)

Rank	Country	2005-2009 Average	2009 Imports	Share of Imports	Imports as a Share of Consumption
1	Canada	3,586	3,271	88%	14%
2	Trinidad & Tobago	356	236	6%	1%
3	Egypt	104	160	4%	1%
4	Norway	9	29	1%	<0.5%
5	Mexico	30	28	1%	<0.5%
6	Nigeria	37	13	<0.5%	<0.5%
7	Qatar	7	13	<0.5%	<0.5%
8	Algeria	38	0	0%	0%
9	Equatorial Guinea	4	0	0%	0%
10	Malaysia	2	0	0%	0%
11	Oman	<1	0	0%	0%
TOTAL		4,163	3,737	100%	16%

Source: EIA's U.S. Natural Gas Imports by Country, http://www.eia.gov/dnav/ng/ ng_move_impc_s1_a.htm.

Notes: *GECF Member*. The United States had imported LNG from Australia, Brunei, Indonesia, and UAE prior to the time period examined in this table.

Table D-2. U.S. Exports of Natural Gas
units = bcf

Rank	Country	2005-2009 Average	2009 Exports	Share of Exports	Exports as a Share of Production
1	Canada	488	701	65%	3%
2	Mexico	324	338	32%	2%
3	Japan	48	31	3%	<0.5%
4	South Korea	1	3	<0.5%	<0.5%
5	Russia	<0.5	0	0%	0%
TOTAL		862	1,073	100%	5%

Source: EIA's U.S. Natural Gas Exports by Country, http://www.eia.gov/dnav/ng/ng_move_expc_s1_m.htm

Notes: The United States has exported natural gas as LNG to both Canada and Mexico in addition to its more traditional pipeline exports. The LNG exports are incorporated into the figures above, but are relatively negligible.

End Notes

[1] Data in this report are 2009 figures from the *BP Statistical Review of World Energy 2010* unless otherwise noted. For global data, BP's Statistical Review is considered an industry standard.

[2] International Finance Corporation, *Environmental, Health, and Safety Guidelines for Thermal Power Plants*, December 19, 2008, p. 8.

[3] Unconventional natural gas refers to natural gas that is not held in traditional porous rock reservoirs like limestone or sandstone, but is trapped in other types of formations. The three most common forms of unconventional natural gas are coalbed methane, shale gas, and tight gas. Coalbed methane refers to natural gas associated with coal seams. Shale gas refers to natural gas trapped in shale rock, which tends to be fine-grained sedimentary rock. Tight gas refers to natural gas trapped in impermeable and non-porous formations.

[4] Resources, or resource base, is a broad term that includes reserves (see below) as well as natural gas less likely to be produced. Resources are not subject to today's technology or price constraints as reserves are and may be produced sometime in the future.

[5] Natural gas is liquefied to make transportation by tanker economical. When natural gas is cooled to -260°F it liquefies and reduces its volume by 1/600th. The liquefied gas is then pumped onto specially designed tankers and shipped to ports to be regasified at specially designed import terminals. Once it is back in its gaseous state, the natural gas is pumped into a pipeline system and is no different from normal natural gas.

[6] Reserves is an industry term to define the likelihood that natural gas resources can be produced using current technology and at today's prices, according to the Society of Petroleum Engineers and the World Petroleum Congresses definition.

[7] U.S. Energy Information Administration (EIA), *Summary: U.S. Crude Oil, Natural Gas, and Natural Gas Liquids Proved Reserves 2009*, November 2010, http://www.eia.gov/pub/oil_gas/natural_gas/ data_publications/ crude_oil_natural_gas_reserves/current/pdf/arrsummary.pdf.

[8] *BP Statistical Review of World Energy 2010*, p. 29.

[9] U.S. Energy Information Administration (EIA), *Natural Gas Consumption by End Use Database*, September 29, 2010.

[10] Reserves is an industry term to define the likelihood that natural gas resources can be produced using current technology and at today's prices according to the Society of Petroleum Engineers and the World Petroleum Congresses definition.

[11] Production has increased along with the addition to reserves, which is why the reserves-to-production ratio has stayed constant.

[12] *BP Statistical Review of World Energy 2010*, p. 22.

[13] U.S. Energy Information Administration (EIA), *Summary: U.S. Crude Oil, Natural Gas, and Natural Gas Liquids Proved Reserves 2009*, November 2010. This value is not used for international comparisons in this report because similar updated values do not exist for other nations.

[14] The Potential Gas Committee (PGC) is an independent, nonprofit organization made up of knowledgeable volunteer members who work in various part of the natural gas industry. PGC is loosely affiliated with the Colorado School of Mines through the school's Potential Gas Agency. Funding for PGC comes mostly from industry-related organizations.

[15] U.S. Energy Information Administration (EIA), *Summary: U.S. Crude Oil, Natural Gas, and Natural Gas Liquids Proved Reserves 2009*, November 2010, pp. 1 and 4.

[16] Hydraulic fracturing is an industry practice of pumping water and proppant, a granular material used to hold open fractures, into wells to improve recovery of natural gas. For additional information on hydraulic fracturing, see CRS Report R40894, *Unconventional Gas Shales: Development, Technology, and Policy Issues*, coordinated by Anthony Andrews.

[17] In its FY2010 Appropriations Committee Conference Report, Congress directed EPA to study the relationship between hydraulic fracturing and drinking water.

[18] EIA tracks gas prices from the wellhead or at the well, which is the lowest price, industrial prices for manufacturing and other uses, commercial prices for nonmanufacturing activities, electric power, and residential use.

[19] International Energy Agency, *World Energy Outlook 2010*, November 2010.

Natural Gas: Outlooks and Opportunities
Editor: Lucas N. Montauban

ISBN: 978-1-61324-132-5
© 2011 Nova Science Publishers, Inc.

Chapter 2

DISPLACING COAL WITH GENERATION FROM EXISTING NATURAL GAS-FIRED POWER PLANTS[*]

Stan Mark Kaplan

SUMMARY

Reducing carbon dioxide emissions from coal plants is a focus of many proposals for cutting greenhouse gas emissions. One option is to replace some coal power with natural gas generation, a relatively low carbon source of electricity, by increasing the power output from currently underutilized natural gas plants.

This report provides an overview of the issues involved in displacing coal-fired generation with electricity from existing natural gas plants. This is a complex subject and the report does not seek to provide definitive answers. The report aims to highlight the key issues that Congress may want to consider in deciding whether to rely on, and encourage, displacement of coal-fired electricity with power from existing natural gas plants.

The report finds that the potential for displacing coal by making greater use of existing gas-fired power plants depends on numerous factors. These include:

- The amount of excess natural gas-fired generating capacity available.
- The current operating patterns of coal and gas plants, and the amount of flexibility power system operators have for changing those patterns.
- Whether or not the transmission grid can deliver power from existing gas power plants to loads currently served by coal plants.
- Whether there is sufficient natural gas supply, and pipeline and gas storage capacity, to deliver large amounts of additional fuel to gas-fired power plants.

There is also the question of the cost of a coal displacement by gas policy, and the impacts of such a policy on the economy, regions, and states.

All of these factors have a time dimension. For example, while existing natural gas power plants may have sufficient excess capacity today to displace a material amount of

---

[*] This is an edited, reformatted and augmented version of a Congressional Research Services publication, dated January 19, 2010.

coal generation, this could change in the future as load grows. Therefore a full analysis of the potential for gas displacement of coal must take into account future conditions, not just a snapshot of the current situation.

As a step toward addressing these questions, Congress may consider chartering a rigorous study of the potential for displacing coal with power from existing gas-fired power plants. Such a study would require sophisticated computer modeling to simulate the operation of the power system to determine whether there is sufficient excess gas fired capacity, and the supporting transmission and other infrastructure, to displace a material volume of coal over the near term. Such a study could help Congress judge whether there is sufficient potential to further explore a policy of replacing coal generation with increased output from existing gas-fired plants.

INTRODUCTION

Purpose and Organization

Coal-fired power plants currently account for about 80% of CO2 emissions from the U.S. electric power industry and about 33% of all U.S. CO2 emissions.[1] Accordingly, reducing CO2 emissions from coal plants is a focus of many proposals for cutting greenhouse gas emissions. Options include capturing and sequestering the CO2 emitted by coal plants, and/or replacing coal-fired generation with low- and zero-carbon sources of electric power, such as wind or nuclear power.

Another option is to replace coal power with increased use of natural gas generation. Natural gas is not a zero-carbon fuel, but gas-fired power using modern generating technology releases less than half of the CO2 per megawatt-hour (MWh) as a coal plant. Recent large increases in estimates of natural gas reserves and resources, especially from shale formations, have further fed interest in natural gas as a relatively low carbon energy option.

One proposal is that the nation can and should achieve near-term reductions in carbon emissions by making more use of existing natural gas plants. This argument was made at an October 2009 Senate Energy and Natural Resources Committee hearing on *The Role of Natural Gas in Mitigating Climate Change*. An executive for a large natural gas pipeline company stated that "Just as natural gas plays a key role in meeting U.S. energy demands, it can also play a key role in providing meaningful, *immediate*, and verifiable [CO2] emission reductions."[2] [emphasis added] The witness for Calpine, a large operator of gas-fired power plants, stated that:

> I am here today to tell you that we could, today, simply through the increased use of existing natural-gas fired power plants, meaningfully reduce the CO₂ emissions of the power sector, *immediately* and for the foreseeable future. *In other words, a near- and medium-term solution to our climate change challenge is at hand.* No guesswork. No huge spending programs needed. That power would be reliable—available all day, every day. And if we embrace this solution with the right incentives, American business would continue to invest its own capital in existing proven technologies to build even more natural gas fired plants to dramatically further reduce emissions for the longer term. [emphasis added][3]

Both of these statements emphasize the claimed *immediate* carbon reductions that can result from increased use of natural gas. This would be accomplished by squeezing more electricity from existing gas-fired power plants, so that coal-fired plants can be operated less and CO_2 emissions quickly and substantially reduced.

This report provides an overview of the issues involved in displacing coal-fired generation with electricity from existing natural gas plants. This is a complex subject and the report does not seek to provide definitive answers. The report aims to highlight the key issues that Congress may consider in deciding whether to rely on, and encourage, displacement of coal-fired electricity with power from existing natural gas plants.

The balance of the report is organized as follows:

- Background on gas-fired generation and capacity.
- Coal displacement feasibility issues.
- Policy considerations.

The report also includes two appendices. Appendix A, Background on the Electric Power System, may be of particular value to readers relatively new to the subject. Appendix B provides information on the gas-burning combined cycle generating technology discussed in the report.

Issues Not Considered in the Report

Several topics are beyond the scope of this report:

- *What would be the cost of a policy of displacing coal with natural gas?* The cost would depend on a host of uncertain variables, such as future natural gas and coal prices, any need to build additional pipeline and transmission line facilities, and the cost of carbon (if any).
- *Could natural gas be burned on a large scale in existing coal plants?* Assessing this option would require engineering analysis of the plants and determining how many coal plants have access to high capacity natural gas pipelines.
- *How will circumstances change over time?* For example, while existing natural gas plants may have enough excess capacity today to displace a material amount of coal generation, this could change in the future as load grows.
- *What kind of existing natural gas plants could be used to displace coal?* This report focuses on the potential for displacing coal generation with increased use of underutilized "combined cycle" generating plants, the most modern and efficient type of natural gas-fired power plants. Two other types of gas-fired plants have low utilization rates: peaking plants (stand-alone combustion turbines and diesel generators) and old steam-electric natural gas plants. These are not reviewed in the report because they are relatively inefficient and may not be designed or permitted for baseload operation.

Addressing these issues would require computer modeling and engineering analysis beyond the scope of this report. As noted in the concluding section of the report, these issues, if of interest to Congress, could be part of a more comprehensive review of the potential for displacing coal with natural gas.

BACKGROUND ON GAS-FIRED GENERATION AND CAPACITY

The argument for displacing coal with natural gas rests on the fact that the United States has a large base of advanced technology, underutilized, gas-burning power plants. This section of the report describes how this reservoir of underutilized natural gas combined cycle (NGCC) plants came about, and why it may represent an option for reducing the use of coal plants.

Capacity Trends

From the 1990s into this century, gas-fired power plants have constituted the vast majority of new generating capacity built in the United States. This development is illustrated by **Figure 1** for the period 1990 to 2007. Minimal new coal capacity was constructed and the growth in nuclear capacity was limited to uprates to existing plants. Only wind capacity has challenged the preeminence of natural gas as the source of new generating capacity, and then only in the latter part of the 2000s when total capacity additions declined sharply.

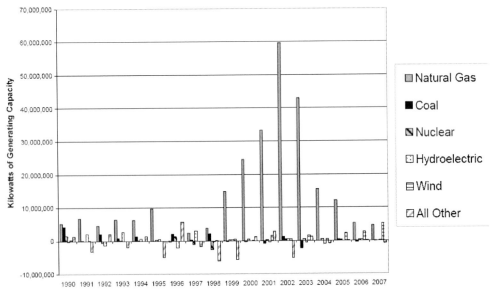

Source: Calculated from data in EIA, *Annual Energy Review 2008*, Table 8.11a, http://www.eia.doe.gov/emeu/aer/ elect.html.

Notes: Capacity can decrease when retirements and deratings of units exceed capacity additions and increases. Also, in some cases the primary fuel of a unit may change, such as from wood to coal. The net change is calculated as the year over year change for each type of capacity.

Figure 1. Net Change in Generating Capacity by Energy Source, 1990 to 2007; Net Summer Capacity.

As shown in **Figure 2**, this building boom doubled the natural gas share of total generating capacity between 1989 and 2007. Natural gas-fired capacity is now the largest component of the national generating fleet.

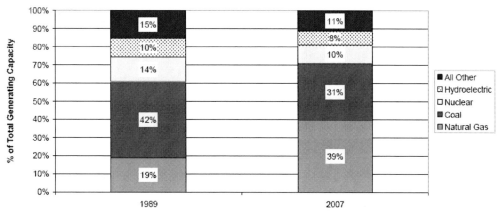

Source: Calculated from data in EIA, *Annual Energy Review 2008*, Table 8.11a, http://www.eia.doe.gov/emeu/aer/ elect.html.

Figure 2. Shares of Total Generating Capacity by Energy Source, 1989 and 2007; Shares of Total Net Summer Capacity.

Although natural gas is the largest source of generating capacity, it trails far behind coal as a source of actual electricity generation.[4] In 2008, coal accounted for 49% of all electricity produced, compared to 21% for natural gas, 20% for nuclear power, and 6% for hydroelectric generation.[5] The remainder of this section will explain why so much new gas-fired generation was built and why it is underutilized.

Factors Supporting the Boom in Gas-Fired Plant Construction

Natural gas was the major source of new capacity in the 1990s and early 2000s in part by default. Nuclear and coal power have been burdened with cost, environmental, and (in the case of nuclear power) safety concerns. Oil-fired generation was essentially ruled out by the costs and supply risks of petroleum supplies. This left natural gas as the energy source for new non-renewable power plants. But in addition to the negatives that surrounded alternatives, gas fired capacity also grew because of favorable technological, cost, environmental, and power market characteristics.

Technology
The new gas-fired plants constructed in the 1990s and subsequently were built around the latest design of combustion turbines—a specialized form of the same kind of technology used in a jet engine, but mounted on the ground and used to rotate a generator. Stand-alone combustion turbines were built to serve as peaking units that would operate only a few hundred hours a year. However, the most important technological development was the application of combustion turbines in modern natural gas combined cycle power plants. (For additional information see Appendix B.) These plants were often intended to serve as

baseload generators which would operate 70% or more of the time. The NGCC has three important characteristics:

- **The technology is very efficient**, because it makes maximum use of the energy in the fuel through a two-step generating process that captures waste heat that would otherwise be lost.[6]
- **NGCC plants can be built relatively quickly and cheaply**. An NGCC plant costs roughly $1,200 per kilowatt of capacity, about half as much as for a coal-fired plant, and can be built in about two to three years from ground-breaking to operation. This compares to about to five to six years to build a coal plant. Coal plants also tend to have longer pre-construction planning and permitting phases.[7]
- **Combined cycle technology is suitable for relatively small scale and modular construction.** NGCC plants can be economically built at unit sizes of about 100 MW, and larger projects can be constructed by adding units in a building block fashion over time. Coal plants in contrast are generally economical only at a unit size of several hundred megawatts.

For the reasons discussed below, these characteristics made the NGCC an attractive technology option for the independent power producers that dominated the construction of new power plants in the 1990s and after.

Natural Gas Prices

The construction of new gas-fired capacity was also encouraged by relatively low natural gas prices in the 1990s. As illustrated by **Figure 3**, the spot price for natural gas hovered around $2.00 to $3.00 per MMBtu (nominal dollars) through the decade, and a widely held expectation was that gas prices would remain low into the future.[8]

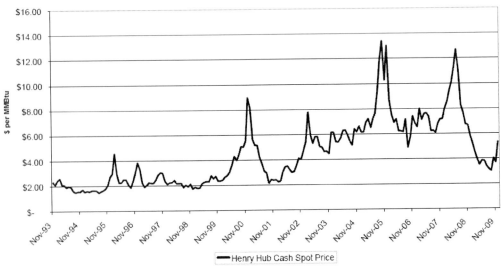

Source: U.S. Federal Reserve Bank of St. Louis FRED database (http://research.stlouisfed.org/fred2/series/ GASPRICE?cid=98).

Figure 3. Henry Hub Cash Spot Price for Natural Gas.

Carbon Dioxide Emissions

The operation of NGCC plants, and natural gas plants generally, produce fewer harmful environmental impacts than coal-fired plants, and have been much easier to site and permit than coal plants. NGCC technology has fewer air emissions than coal plants in part because of the nature of the fuel, and in part because of the greater efficiency of the technology. For example, natural gas when burned inherently emits about half as much carbon dioxide as coal.[9] However, because combined cycle plants are more efficient than typical existing coal plants in converting fuel into electricity, the difference in emissions is greater when measured in terms of CO_2 released per megawatt-hour of electricity produced. By this measure a modern combined cycle emits only about 40% of the CO_2 per MWh as a typical existing coal plant.[10]

Electric Power Industry Restructuring and Overbuilding

Restructuring of the electric power industry (beginning in the late 1970s and accelerating in the 1990s) included federal and state policies that encouraged the separation of power generation and power plant construction from other utility functions. In the 1990s, new independent power producers (IPPs) bought power plants from utilities and constructed most of the new generating capacity. Because these companies sold power into competitive markets and did not have the security of regulated rates and guaranteed markets, they generally sought to minimize risks by constructing relatively low cost, quick-to-build, power plants.

For these reasons, independent power producers built many NGCC plants, largely to meet baseload demand. As shown in **Table 1**, between 1990 and 2007 over 168,000 MW of NGCC capacity was built at 345 plant sites. This was an enormous building program, equivalent to adding 23% to the entire national generating fleet that existed in 1990. However, the growth in generating capacity did not stop with new combined cycle plants. As also shown in **Table 1**, another 89,843 MW of less efficient stand-alone peaking turbines were constructed, plus another 56,939 MW of other generating technologies. When all of this capacity is added together, generating capacity grew by 43% between 1990 and 2007.

Table 1. Growth in Generating Capacity, 1990 – 2007

	Natural Gas Combined Cycle (NGCC)	Stand-Alone Combustion Turbine (Natural	All Other Fuel Sources and Technologies	Total
Additions to Generating Capacity, 1990 –2007 (MW)	168,259	89,843	56,939	315,041
Additions as a Percent of Total 1990 Generating Capacity (734,100 MW)	23%	12%	8%	43%

Source: Calculated from the EIA-860 data file for 2007, http://www.eia.doe.gov/cneaf/electricity/page/eia860.html.

Notes: The capacity shown is net summer capacity. The All Other category contains 2,184 MW of gas-fired capacity, primarily in new internal combustion engines and steam turbines, and 5,432 MW of combined cycle and stand-alone combustion turbine capacity burning fuels other than natural gas (primarily stand-alone combustion turbines using distillate fuel oil).

By the mid-2000s it was apparent that the combined cycle building boom had resulted in excess and underutilized generating capacity. Too many plants were built, in part because of questionable investment decisions by independent developers operating in an immature restructured power market. The capacity glut was compounded by a dramatic increase in gas prices after 2000. (See **Figure 3**.) Even the high efficiency of the combined cycle plants could not compensate for gas prices that at times peaked above $10.00 per MMBtu, compared to $2.00 to $3.00 per MMBtu prices (nominal dollars) in the 1990s.

The consequence of the combined cycle building boom and bust is that the fleet of NGCC plants has a large amount of unused generating capacity, as illustrated in **Table 2** for a "study group" of large combined cycle plants defined for this report.[11] Baseload operation can be reasonably defined as operation at an annual capacity factor of 70% or greater. As shown in **Table 2**, only 13% of combined cycle capacity in the study group operated in this range in 2007. A third of the combined cycle capacity had a utilization rate of less than 30%; that is, the plants were the equivalent of idle more than 70% of the time.

Table 2. Utilization of Study Group NGCC Plants, 2007

Capacity Factor Category	Net Summer Megawatts	Percent of Total NGCC Megawatts	Number of NGCC Plants	Percent of Total NGCC Plants
70% and Greater	22,151	13%	42	13%
Under 70% to 50%	40,103	24%	68	22%
Under 50% to 30%	50,711	30%	90	29%
Under 30%	57,662	34%	114	36%
Total	170,627	100%	314	100%

Source: Calculated from the EIA-860 and EIA-906/920 databases for 2007 (http://www.eia.doe.gov/cneaf/ electricity/page/data.html).

Notes: Detail many not add to total due to rounding. For information on the characteristics of the power plants selected for the study group, see footnote 11.

In 2007 the study group of NGCC plants had an average capacity factor of 42%.[12] In contrast, the study group of coal plants had an average capacity factor of 75%.[13] It is this mismatch between combined cycle and coal plant operating patterns—the former, low carbon emitting but underutilized; the latter, high carbon emitting and highly utilized—that creates the interest and perceived opportunity for displacing coal power with gas generation from existing plants.

COAL DISPLACEMENT FEASIBILITY ISSUES

Estimates of Displaceable Coal-Fired Generation and Emissions

The maximum coal-fired generation and emissions that may be displaceable by existing NGCC plants is estimated in Table 3 and Table 4. As noted above, the plants in the NGCC study group had an average capacity factor of 42% in 2007. As shown in the tables, if the utilization of this capacity could be essentially doubled to 85%, it would generate additional power equivalent to 32% of all coal-fired generation in 2007, and could displace about 19% of the CO2 emissions associated with coal-fired generation of electricity.

**Table 3. Approximation of the Maximum Displaceable
Coal-Fired Generation, Based on 2007 Data**

(1)	(2)	(3)	(4)	(5)
Actual NGCC Generation, 2007 (MWh)	Hypothetical NGCC Generation at an 85% Capacity Factor (MWh)	Hypothetical Surplus Generation Available for Coal Displacement (MWh)	Actual Coal-Fired Generation in 2007 (MWh)	Hypothetical Surplus NGCC Generation as a Percentage of Coal Generation
		(2) – (1)		(3)/ (4)
630,358,373	1,270,487,153	640,128,780	2,016,456,000	32%

Source: CRS estimates based on EIA-906/920 and EIA-860 electric power databases, and EIA, *Electric Power Annual 2007*, Table ES1, http://www.eia.doe.gov/cneaf/ electricity/ epa/ epa_ sum.html.

Notes: The generation in column 1 is for the 314 NGCC plants included in the study group defined for this report. For additional information see footnote 11. As discussed in the main body of the report, several factors, such as transmission system limitations, will tend to drive actual displacement below the maximum potential. Also, this estimate is for 2007, and in other years the amount of surplus gas generation and the amount of coal generation will likely vary from 2007 values.

**Table 4. Approximation of Maximum Displaceable CO_2
Emitted by Coal-Fired Generators, Based on 2007 Data**

(1)	(2)	(3)	(4)	(5)	(6)
Estimated Hypothetical Coal Generation Displaced by Natural Gas (MWh)	Estimated CO2 Emissions from Displaced Coal Generation (Million Metric Tons)	Estimated CO2 Emissions From NGCC Generation Used to Displace Coal (Million Metric Tons)	Net Reduction in Emissions of CO2 by Natural Gas Displacement of Coal (Million Metric Tons)	Total CO2 Emissions from Coal for Power Generation, 2007 (Million Metric Tons)	Hypothetical Net Reduction in CO_2 Emissions as a Percentage of 2007 Total Electric Power Coal Emissions of CO_2
			(2) – (3)		(4) / (5)
640,128,780[a]	635.7[b]	253.6[c]	382.1	2,002.4	19%

Source: CRS estimates based on: EIA-906/920 database (http://www.eia.doe.gov/cneaf/ electricity/page/ eia906_920.html); EIA, *Electric Power Annual 2007*, Table A3, http://www.eia.doe.gov/cneaf/electricity/epa/ epa_sum.html; EIA, *Annual Energy Review 2008*, Table 12.7a, http://www.eia.doe.gov/emeu/aer/envir.html.

Notes: As discussed in the main body of the report, several factors, such as transmission system limitations, will tend to drive actual displacement below the maximum potential. Also, this estimate is for 2007, and in other years the amount of surplus gas generation and the amount of coal generation will likely vary from 2007 values.

[a] From Table 3, column 3.

[b] In 2007, total coal generation was 2,016,456,000 MWh (Table 3, column 4) and total CO_2 emissions from coal were 2,002.35 million metric tons. This equates to 0.993 metric tons of CO_2 per MWh of coal generation (the comparable value for a modern NGCC plant is about 0.4 metric tons of CO_2

per MWh). Therefore, the estimated CO₂ emissions from the displaced coal is 640,128,780 MWh x 0.993005 metric tons of CO₂ per MWh = 635.651 million metric tons of CO₂.

[c] Actual study group NGCC generation in 2007 was 630,358 MWh (Table 3, column 1). This generation consumed 4,702,226,931 MMBtus of natural gas, or 7.4596 MMBtus of gas per MWh. At this average heat rate, it would take 4,775,104,647 MMBtus of gas to displace 640,128,780 MWh of coal generation. This much gas burn would release 253.6 million metric tons of CO₂, using an emissions factor of 117.08 pounds of CO₂ per MMBtu of natural gas consumed and 2,204.6 pounds per metric ton.

Although these calculations suggest that at most about a third of current (2007) coal-fired generation could be displaced by existing NGCC plants, it is unlikely that this maximum could actually be achieved. This section of the report will discuss issues that relate to the feasibility of actually displacing coal with gas from existing power plants. The issues are:

- Transmission system factors;
- System dispatch factors;
- Natural gas supply and price; and
- Natural gas transportation and storage.

Transmission System Factors

If an NGCC generating unit is located at the same plant site as a coal-fired generating unit, it is probably fair to assume that the NGCC unit can use the same transmission lines as the coal unit and can transmit its power to any load the coal unit is used to meet. However, in most cases coal units and NGCC units are built at separate locations and rely on different transmission paths to move their power. This means that there is no guarantee that the NGCC plant can send its power to the same loads as the coal plant and by doing so displace coal-fired generation.

Even on a regional level, coal and NGCC plants are not necessarily located in the same areas. The maps in **Figure 4** and **Figure 5** show, respectively, the location of large coal and NGCC plants in the conterminous states. The maps show that in some cases coal and NGCC plants are in the same regions, such as east Texas. On the other hand, California has many NGCC plants and no coal plants, while the Ohio River valley has a dense concentration of coal plants and only a handful of NGCC plants.

This section of the report will discuss three types of transmission system constraints that can prevent one power plant from meeting the load currently served by another plant. These limits on the "transmission interchangeability" of coal and NGCC plants are:

- Isolation of the Interconnections;
- Limited long-distance transmission capacity; and
- Transmission system congestion.

The concluding part of this section presents an analysis of potential coal displacement by gas using the proximity of coal and existing NGCC plants as a proxy for transmission interchangeability.

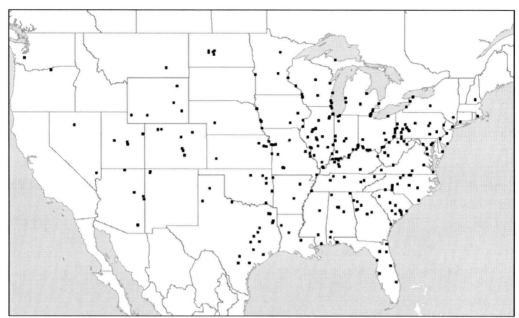

Source: Platts Powermap (fourth quarter 2009 database).

Figure 4. Location of Large Coal-Fired Power Plants in the Conterminous States; 250 Megawatt and Greater Net Summer Capacity.

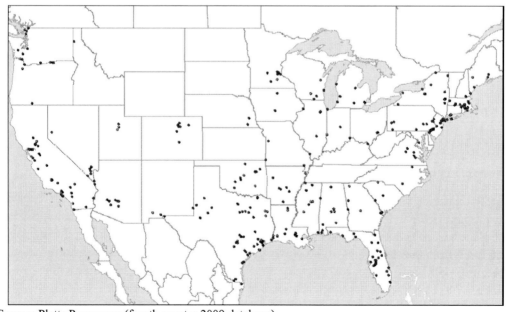

Source: Platts Powermap (fourth quarter 2009 database).

Figure 5. Location of Large NGCC Power Plants in the Conterminous States; 100 Megawatt and Greater Net Summer Capacity.

The concluding part of this section presents an analysis of potential coal displacement by gas using the proximity of coal and existing NGCC plants as a proxy for transmission interchangeability.

Isolation of the Interconnections

The electric power grid covering the conterminous states is divided into three "interconnections," Eastern, Western, and the ERCOT Interconnection that covers most of Texas (Figure 6). These three interconnections operate in most respects as independent systems. There are only a handful of limited, low capacity links between the interconnections. Consequently, surplus capacity in one interconnection cannot be used to meet load in another interconnection.

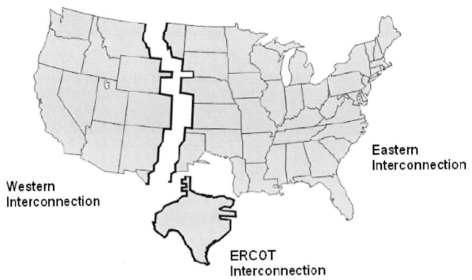

Source: adapted from a map located on the Energy Information Administration website at http://www.eia.doe.gov/cneaf/electricity/page/fact_sheets/transmission.html.
Notes: ERCOT = Electric Reliability Council of Texas.

Figure 6. United States Power System Interconnections.

To illustrate with a hypothetical example, assume 1,000 MW of surplus NGCC capacity in the northern part of the ERCOT Interconnection, and a desired use for that capacity to displace coal in Oklahoma, which is in the Eastern Interconnection. Although the regions are adjacent, from the standpoint of the power grid they are electrically isolated from each other because, with very limited exceptions, the ERCOT and Eastern Interconnections are not linked. Therefore the displacement cannot take place.

Limited Long-Distance Transmission Capacity

Within each interconnection the network of power lines, generating plants, and electricity consumers are linked together. The grid operates in some respects like a single giant machine in which, for example, a disturbance in the operation of the transmission system in Maine is detectible by system monitors in Florida.

Although all generators and loads within an interconnection are linked by the grid, the power grid is not designed to move large amounts of power long distances. The grid was not built in accordance with a "master plan," analogous to the Interstate Highway System. Transmission lines were first built in the early 20th century by single utilities to move electricity to population centers from nearby power plants. As generation and transmission technology advanced, the distances between power plants and loads increased, but the model of a single entity building lines within its service territory to supply its own load still predominated.

Over time the local grids began to interconnect, due to utilities building jointly owned power plants and because power companies began to grasp the economic and reliability benefits of being able to exchange power. Nonetheless, this pattern of development did not emphasize the construction of very long-distance inter-regional lines. Consequently, the capacity to move power long distances within interconnections is limited. For example, while a generator in Maine and a load in Florida are connected by the grid, it is not feasible to send power from Maine to Florida because the transmission lines do not have enough capacity to move the electricity.

Additionally, over distances of hundreds of miles, losses occur with transmission of electricity, making the transfer uneconomic.[14] Power can be moved long distances most efficiently by the highest voltage transmission lines, but only a small portion of the national grid consists of these types of lines.[15] Much of the debate over the proposed increased use of renewable power involves how to build and pay for the new transmission lines that would be needed to move wind and solar power from remote locations to population centers, in part to displace fossil-fueled power plants. Coal displacement by existing gas-fired generators is a similar type of problem. If the existing transmission network does not have sufficient capacity in the right places, then it may not be practical to move gas-powered electricity to loads currently served by coal plants without investing in upgraded or new power lines.

Transmission System Congestion

Even across relatively short distances, options for moving power can be restricted by transmission line congestion. Transmission congestion occurs when use of a power line is limited to prevent overloading that can lead to failure of the line. Congestion can occur throughout a power system:

- Regional congestion: for example, power flows are limited between the eastern and western parts of the PJM power pool (covering much of the Midwest and middle Atlantic regions) by congestion.[16] In the western states, examples of congested links include power flows between Montana and the Pacific Northwest, and between Utah and Nevada.[17]
- State-level congestion: for example, congestion restricts power flows into and out of southwestern Connecticut.[18]
- Local congestion: These are "load pockets" with limited ability to import power. New York City is an example of a load pocket.

Transmission congestion can increase costs to consumers by forcing utilities to depend on nearby inefficient power plants to meet load instead of importing power from more distant but less costly units. Studies suggest that the annual costs of transmission congestion range from the hundreds of millions to billions of dollars.[19] However, for the purposes of this report the key aspect of transmission system congestion is not the cost impact, but the restrictions it imposes on power flows. Because of congestion, it may not be possible to ship power from an underutilized NGCC plant to a load served by coal power, because the transmission path available to the combined cycle is too congested to carry the electricity.

The solution for congestion is not necessarily massive transmission construction. For example, DOE found that in the Eastern Interconnection "a relatively small portion of constrained transmission capacity causes the bulk of the congestion cost that is passed through to consumers. This means that a relatively small number of selective additions to transmission capacity could lead to major economic benefits for many consumers."[20] However, in the absence of this construction, congestion remains a constraint on the choice of power plants available to meet a load.

Power Plant Proximity Analysis

Transmission system limitations on coal displacement can be rigorously analyzed using sophisticated computer models. Such an analysis is beyond the scope of this report. However, a first approach to the significance of transmission factors can be made by examining how close coal plants are to existing NGCC plants. The assumption behind such a "proximity analysis" is that the closer an NGCC plant is to a coal plant, the more likely that the NGCC plant will connect to the same transmission lines as the coal plant. If the NGCC plant has this comparable transmission access—that is, the combined cycle is "transmission interchangeable" with the coal plant—it potentially could serve the same load as the coal plant and supplant the coal generation.

CRS performed a proximity analysis for the coal plants and NGCC plants in the study groups defined for this report. The analysis was conducted as follows, in all cases using 2007 data (the most recent pre-recession year for which complete data were available):

1) Study groups of large coal plants and NGCC plants were defined. The plants in these groups accounted for the great majority of power plant coal generation and NGCC generation in 2007.[21]
2) The latitude and longitude of each plant (provided by EIA) was entered into a geographical information system (GIS).
3) The GIS was used to identify all coal plants with one or more existing NGCC plants within a ten mile radius. The hypothetical surplus generation for each NGCC plant within the ten-mile radius was calculated and assumed to displace generation from the coal plant.[22] If one NGCC plant was within ten miles of two or more coal plants, it was allocated first to the coal plant with the largest estimated CO2 emissions in 2007.[23]
4) A second version of Step 3 was performed which included all NGCC plants within 25 miles of a coal plant.

The maps in **Figure 7** and **Figure 8** show the locations of the coal plants assumed to have generation displaced by existing NGCC plants.[24]

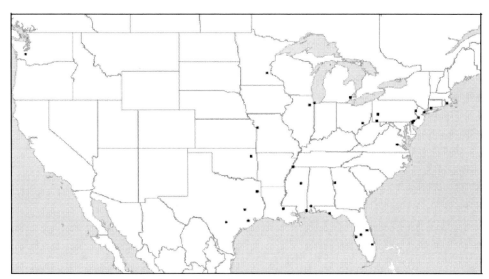

Source: CRS estimates, mapped using the Platts Powermap system.
Notes: The maps only show coal plants assumed to have had generation displaced. If a coal plant has an NGCC plant with the ten mile radius, but the NGCC plant was assumed to be unavailable to displace coal (for example, because it had a capacity factor in 2007 of 85% or higher) the coal plant is not shown on the map.

Figure 7. Coal Plants with Hypothetical Generation Displaced by a NGCC Plant Within 10 Miles.

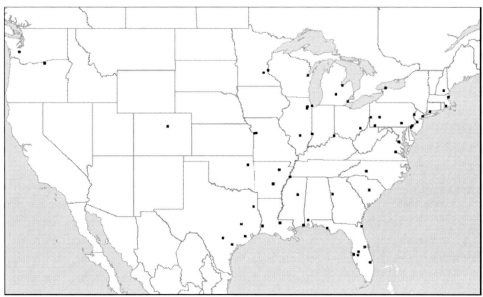

Source: CRS estimates, mapped using the Platts Powermap system.
Notes: The maps only show coal plants assumed to have had generation displaced. If a coal plant has an NGCC plant with the 25 mile radius, but the NGCC plant was assumed to be unavailable to displace coal (for example, because it had a capacity factor in 2007 of 85% or higher) the coal plant is not shown on the map.

Figure 8. Coal Plants with Hypothetical Generation Displaced by a NGCC Plant Within 25 Miles.

This analysis is not a forecast. It is a first approach to estimating coal displacement potential based on one factor, the proximity of coal and existing NGCC plants. Many other factors, including, for example, how utility systems are dispatched, the configuration and capacity of the electric power transmission system, fuel cost and availability, natural gas transportation capacity, and power system reliability requirements, would influence actual coal displacement potential. These other factors could increase or decrease the potential displacement.

Table 5, which gives the results of the proximity analysis, shows in column 4 that existing NGCC plants located near coal plants might be able to achieve 15% to 28% of the potential maximum coal generation and CO_2 emissions displacement. However, the *displaceable* coal generation and emissions (see **Table 3** and **Table 4**) are only a fraction of *total* U.S. coal generation and CO_2. As shown in **Table 5**, column 5, the hypothetical displaced coal generation and emissions are equivalent to 5% to 9% of *total* U.S. coal generation, and 3% to 5% of the associated CO_2 emissions.

Given its limitations, the analysis suggests that existing NGCC plants near coal plants may be able to account for something on the order of 30% or less of the displaceable coal-fired generation and CO_2 emissions. Greater displacement of coal by existing NGCC plants would depend on more distant NGCC plants which would be less clearly "transmission interchangeable" with coal plants. This emphasizes the importance that the configuration and capacity of the transmission system will likely play in determining the actual potential for displacing coal with power from existing NGCC plants.

Table 5. Hypothetical Estimates of the Displacement of Coal Generation and Emissions by Existing NGCC Plants Based on Proximity Based on 2007 Data

Case	Category	Amount Displaced	Amount Displaced as a % of the Maximum Potential Displacement of Coal by Existing NGCC Plants[a]	Amount Displaced as a % of Total Electric Power Sector Coal MWh and Associated CO_2 Emissions
(1)	(2)	(3)	(4)	(5)
Generation and CO2 Displaced for Coal Plants within 10 Miles of a NGCC Plant	Generation	101.8 Million MWh	16%	5%
	CO2 Emissions	58.1 Million Metric Tons	15%	3%
Generation and CO2 Displaced for Coal Plants within 25 Miles of a NGCC Plant	Generation	181.5 Million MWh	28%	9%
	CO2 Emissions	104.8 Million Metric Tons	27%	5%

Source: CRS estimates primarily based on EIA data. See the main text of the report for more information. For detailed backup, such as lists of plants, contact the author.

Notes: This is not a fo proximity of coal and existing NGCC plants. Many other factors, including, for example, how utility systems are dispatched, the configuration and capacity of the electric power transmission system, fuel cost and availability, natural gas transportation capacity, and power system reliability requirements, would influence actual coal displacement potential. These other factors could increase or decrease the potential displacement. MWh =Megawatt-hours; NGCC = natural gas combined cycle.

a. The values in this column are calculated using column 3; Table 3, column 3; and Table 4, column 4.
b. The values in this column are calculated using column 3; Table 3, column 4; and Table 4, column 5.
c. The study group included 298 coal-fired plants. In the ten-mile radius case, coal is displaced in whole or part at 35 of these plants (11.7% of the plants). In the 25-mile radius case, coal is displaced in whole or part at 60 of these plants (20.1%).

System Dispatch Factors

System dispatch refers to the pattern in which power plants are turned on and off, and their power output ramped up and down, to meet changing load patterns. (For additional discussion, see **Appendix A**.) The concept of displacing coal generation with power from existing NGCC plants assumes that the NGCC plants are underutilized or idle when coal plants are operating. However, this is not necessarily the case. This can be illustrated by examining the monthly utilization of the coal and gas-fired plants in the study groups (**Figure 9**). As shown in the figure, the utilization of coal and combined cycle plants follows a similar pattern: utilization is highest in the summer and, to a lesser degree, in the winter, and lowest in the "shoulder" months of the spring and fall. The figure illustrates that when coal plant operation is at its highest and the most coal power can be displaced, NGCC plant operation is also at its highest and surplus gas-fired generation is therefore at its lowest.

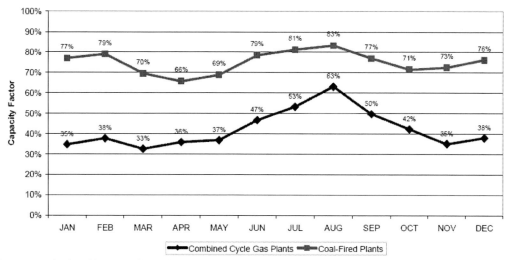

Source: Calculated by CRS from the EIA-906/920 and EIA-860 databases.
Notes: For information on the study groups of coal and NGCC plants, see footnote 11. NGCC= natural gas combined cycle.

Figure 9. Monthly Capacity Factors in 2007 for Study Group Coal and NGCC Plants.

Figure 9 is a national, monthly picture of power plant dispatch. System dispatch actually takes place moment-to-moment, and at this level of detail the complexities in displacing coal with gas become further evident. **Figure 10** graphically illustrates *hourly* dispatch at Plant

Barry, a power plant in Alabama that has both coal and NGCC units at the same site. The data is for November 2007, the month in which the NGCC units at Plant Barry had their lowest generation for the year and therefore, in principle, the most excess capacity available to displace coal.[25] However, the graphic illustrates that even during this low utilization month for the NGCC units, there are still periods when the units were running near maximum output[26] (e.g., November 6 to 9, and 27 to 30). While there were periods when coal plant output was high and the NGCC units were shut down (e.g., November 4), creating the maximum opportunity to displace coal with gas, there were also periods when the NGCC units were available but potential coal displacement was reduced by limited operation of the coal units (e.g., November 18). These examples illustrate the level of detailed analysis required to realistically estimate the potential for changing plant dispatch to displace coal with natural gas.

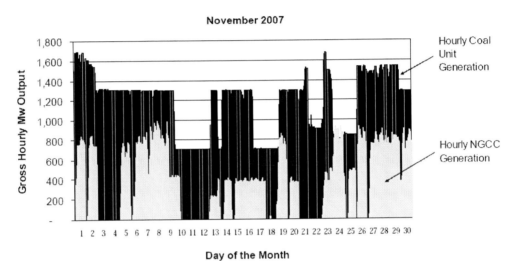

Source: Data downloaded from the EPA website at http://camddataandmaps.epa.gov/gdm/index.cfm?fuseaction=emissions.prepackaged_select.

Figure 10. Hourly Coal and Combined Cycle Generation at Plant Barry.

Natural Gas Supply and Price

Large scale displacement of coal-fired generation by existing NGCC plants could result in a significant increase in U.S. gas demand. **Table 6** compares the actual demand for natural gas for all purposes in 2007 with an illustrative estimate of the additional gas supplies needed if all of the displaceable coal-fired generation (see **Table 3**) were actually replaced by existing NGCC plants. As discussed above, this maximum displacement of coal by existing NGCC plants may be unachievable, so results are also shown for a half and a quarter of the maximum.

Table 6. Illustrative Estimates of Increased Natural Gas Demand For Coal Displacement Compared to Total National Demand Based on 2007 Data

	Hypothetical Maximum Displacement of Coal by Existing NGCC Plants (2007 data and 85% capacity factor)	Half of Hypothetical Maximum Displacement	One Quarter of Hypothetical Maximum Displacement
1. Additional MWh of NGCC Generation Needed to Displace Coal	640,128,780	320,064,390	160,032,195
2. Required Additional Natural Gas in Trillion Btus (Tbtus)	4,775	2,388	1,194
3. Additional Gas as a Percentage of 2007 Gas Consumed for All Purposes in the U.S.[a]	20%	10%	5%

Source: Table 3 and EIA, *Annual Energy Review 2008*, Tables 6.1 and A4 (http://www.eia.doe.gov/emeu/aer/contents.html).

Notes: Total gas consumption in 2007 of 23,047 billion cubic feet was converted to TBtus using a conversion factor of 1.028 (see Table A4 in the *Annual Energy Review*, cited immediately above). The MWh of additional gasfired generation was converted to TBtus using a heat rate of 7.4596 MMBtus of fuel input per MWh. This is the 2007 average annual heat rate for the study group of 314 combined cycle plants, calculated using the generation and fuel input reported in the EIA 906/920 database (http://www.eia.doe.gov/cneaf/electricity/page/eia906_920.html).

[a] Total gas consumption in 2007 for all residential, commercial, industrial, electric power, and transportation purposes was 23,692 Tbtus. The percentages shown in Line 3 are calculated by dividing this number into the values shown on line 2.

Total U.S. natural gas demand in 2007 was the third highest on record. The illustrative estimates of increased gas demand for coal displacement would increase the already high level of demand in 2007 by another 5% to 20% (**Table 6**, line 3).

This increased demand might be met with a combination of increased domestic production, pipeline imports from Canada, Alaskan supplies if the trans-Alaskan gas pipeline is built, and imports of liquefied natural gas by tanker from overseas. For example, one reason for the interest in coal displacement by gas is the recent increase in natural gas available from shale formations and other "unconventional" sources of gas.[27] The combination of higher production (up a projected 3.7% for 2009) and reduced demand due to the 2008-2009 recession has contributed to a sharp decline in gas prices from the peaks experienced earlier in the 2000s (see **Figure 3**).[28] For the longer term, there is widespread optimism concerning the gas supply and price outlook. An example is a late 2009 assessment by the Federal Energy Regulatory Commission (FERC):

> The long-term [gas production] story is one of abundance. In June, the Potential Gas Committee, an independent group that develops biennial assessments of gas resources, raised its estimate to over 2 quadrillion cubic feet, one-third more than its previous level and almost 100 years of gas production at current consumption levels. The large increase is almost entirely due to improvements in our ability to harvest gas from shale and get it to markets at a reasonable cost.... As we have indicated before, gas production is becoming more like mining and manufacturing with high probability of production from

each well drilled. This environment should have profound effects on the traditional boom and bust cycle of gas production.[29]

EIA's most recent long-term forecasts of natural gas wellhead prices for 2020 and 2030 have dropped, respectively, 13% and 11% from its prior forecast, "due to a more rapid ramping up of shale gas production, particularly after 2015. [The forecast] assumes a larger resource base for natural gas, based on a reevaluation of shale gas and other resources...."[30]

Even with the current optimism concerning natural gas supplies and prices, it is important to note that natural gas markets have historically been exceptionally difficult to forecast. According to an EIA self-assessment of its long-term projections, "The fuel with the largest difference between the projections and actual data has generally been natural gas."[31] In the 1990s gas prices were expected to be low; by 2004 prices were much higher than expected and major gas buyers were reported to be "increasingly critical of the nation's system for forecasting natural gas supply and demand."[32] Subsequently, as shown in **Figure 3**, prices plummeted. In the October 2009 Senate hearing on natural gas, a cautionary note was sounded by the witness for Dow Chemical Company:

> Although increased supply from shale gas appears to have changed the production profile, we have seen similar scenarios occur after past spikes. In 1998, significant new imports from Canada came on line; in 2002-2003, there were new supplies from the Gulf of Mexico and in 2005, new discoveries in the Rockies were brought into play. In each case, the initial hopes were too high and production increases were not as large as initially expected.[33]
>
> In 2009, as in 2002, 2004 and 2006, drilling has declined dramatically as price has fallen. After each trough, natural gas demand and price rise once the economy turns, signaling the production community to increase drilling. During the lag between the pricing signals and new production, only one mechanism exists to rebalance supply and demand: demand destruction brought about by price spikes. Demand destruction is an antiseptic economic term for job destruction.[34]

Although multiple options may exist to meet the additional natural gas demand created by a coal displacement policy, the significance of the potential increase in demand should not be underestimated. The lowest level of increased gas demand shown in **Table 6**, 1,194 trillion Btus (TBtus), would raise total demand to 24,886 TBtus. In its most recent Reference Case forecast, the U.S. Energy Information Administration (EIA) does not envision this demand level being reached until after 2028. The middle estimate of increased gas demand shown in **Table 6** would raise total gas demand to 26,080 TBtus, which is larger than EIA's forecast for 2035.[35] A policy of rapid change from coal to gas could therefore involve a significant acceleration of gas demand growth compared to EIA's current estimates.

Natural Gas Transportation and Storage

Gas-fired power plants and other gas consumers receive fuel through a vast national pipeline network. At the end of 2008 the network consisted of 293,000 miles of interstate and intrastate pipelines with the capacity to move up to 215 billion cubic feet (BCF) of gas daily.[36] The capacity of this system is sized to meet peak loads, such as during the winter residential heating season. Peak demands are also supported by a system of natural gas storage facilities connected to the pipeline network. These storage facilities hold gas which is produced during lower demand periods until it is needed to meet peak demand.

It seems unlikely that on a national, aggregate scale, pipeline capacity would be a constraint on coal displacement by existing NGCC plants. The natural gas consumption required for the maximum potential coal displacement by existing NGCC plants (see Table 3) equate to about 15 BCF per day of natural gas, or about 7% of existing pipeline capacity.[37] A 7% increase in peak demand would appear manageable given the planned expansions to the pipeline system (see below). But irrespective of national system-wide capacity, a different question is whether increased use of gas-fired plants could overstress the specific pipelines and storage facilities that serve those plants. This may be an important issue because the increase in gas demand from existing NGCC plants for coal displacement could be large relative to the amount of gas currently used for power generation. As shown in Table 7, illustrative estimates of this increase range from 16% to 66%, which means that the facilities serving those plants could have to handle a material increase in gas demand.

A balancing factor is that the natural gas industry has been effective at adding large amounts of capacity to the pipeline system. Capacity additions in 2007 and 2008 were, respectively, 14.9 and 44.6 BCF per day, and as of mid-2009, 31.9 BCF per day was under construction or approved for construction and completion in 2009. Another 62.1 BCF per day of capacity additions are planned for 2010 and 2011,[38] which is equivalent to almost 30% of current capacity. It appears that, given sufficient lead time, the natural gas industry has the ability to install large amounts of additional transportation capacity to meet increased demand.

Table 7. Illustrative Estimates of Increased Natural Gas Demand Relative to Electric Power Demand, Based on 2007 Data

	Hypothetical Maximum Displacement of Coal by Existing NGCC Plants (2007 data and 85% capacity factor)	Half of Hypothetical Maximum Displacement	One Quarter of Hypothetical Maximum Displacement
1. Additional MWh of Existing NGCC Generation	640,128,780	320,064,390	160,032,195
2. Required Additional Natural Gas in Trillion Btus (Tbtus)	4,775	2,388	1,194
3. Required Additional Gas as a Percentage of Actual Gas Used for Power Generation in 2007[a]	66%	33%	16%

Source: Table 6 and EIA, *Annual Energy Review 2008*, Tables 8.4a (http://www.eia.doe.gov/emeu/aer/contents.html).

Notes: The MWh of additional gas-fired generation was converted to TBtus using a heat rate of 7.4596 MMBtus of fuel input per MWh. This is the 2007 average annual heat rate for the study group of 314 combined cycle plants, calculated using the generation and fuel input reported in the EIA 906/920 database (http://www.eia.doe.gov/cneaf/electricity/page/eia906_920.html).

a. Electric power gas consumption in 2007 was 7,288 Tbtus. The percentages shown in Line 3 are calculated by dividing this number into the values shown on line 2.

POLICY CONSIDERATIONS

As discussed in this report, the potential for displacing coal consumption in the power sector by making greater use of existing NGCC power plants depends on numerous factors. These include:

- The amount of excess NGCC generating capacity available;
- The current operating patterns of coal and NGCC plants, and the amount of flexibility power system operators have for changing those patterns;
- Whether or not the transmission grid can deliver power from existing NGCC plants to loads currently served by coal plants; and
- Whether there is sufficient natural gas supply, and pipeline and gas storage capacity, to deliver large amounts of additional fuel to gas-fired power plants; and consideration of the environmental impacts of increasing gas production.

All of these factors have a time dimension. For example, while existing NGCC plants may have sufficient excess capacity today to displace a material amount of coal generation, this could change in the future as load grows. Therefore a full analysis of the potential for gas displacement of coal must take into account future conditions, not just a snapshot of the current situation.

There is also the question of cost which, as discussed in the introduction, is beyond the scope of this report. Clearly, the cost of a coal displacement by gas policy is highly uncertain, and depends on such factors as future natural gas and coal prices, any need to build additional pipeline and transmission line facilities, and the cost of carbon (if any). The economic impacts of a coal displacement by gas policy could also spill over to other parts of the economy. For example, increased power sector demand could drive up the price of natural gas, to the detriment of other residential, commercial, and industrial users. Decreased production of coal and increased production of natural gas would pose varying costs and benefits for states and regions.

As a step toward addressing these questions, Congress may consider chartering a rigorous study of the potential for displacing coal with power from existing gas-fired power plants. Such a study would require sophisticated computer modeling to simulate the operation of the power system, to determine whether there is sufficient excess gas fired capacity and the supporting transmission and other infrastructure to displace a significant volume of coal over the near term. This kind of study might also estimate the direct costs of a gas for coal policy, such as the impact on electric rates. Because of the large number of uncertainties, such as the future price of natural gas, the study would have to consider several scenarios. Such a study could help Congress judge whether there is sufficient potential to further explore a policy of replacing coal generation with increased output from existing gas-fired plants.

Congress may also consider chartering an analysis of the potential for directly using gas in existing coal-fired plants, either as a supplemental or primary fuel. As noted in the introduction, large scale use of gas in coal plants raises engineering issues and the question of how many coal plants have adequate pipeline connections. However, burning gas in coal plants would make it possible to displace coal while still using existing transmission lines to meet load, which could be a significant advantage.[39]

APPENDIX A. BACKGROUND ON THE ELECTRIC POWER SYSTEM

This appendix provides background on the components and operation of the electric power system. Readers familiar with these topics may wish to skim or skip this appendix.

Power Plants and Power Lines

Power plants, transmission systems, and distribution systems constitute the major components of the existing electric power system, as briefly described and illustrated below (Figure A-1):

- *Generating plants* produce electricity, using either combustible fuels such as coal, natural gas, and biomass; or non-combustible energy sources such as wind, solar energy, or nuclear fuel.
- *Transmission lines* carry electricity from power plants to demand centers. The higher the voltage of a transmission line the more power it can carry and the fewer the line losses during transmission. Current policy discussions focus on the high voltage network (230 kilovolts (kV) rating and greater) used to move large amounts of power long distances.
- Near customers a step-down transformer reduces voltage so the power can be carried by low voltage *distribution lines* for final delivery.

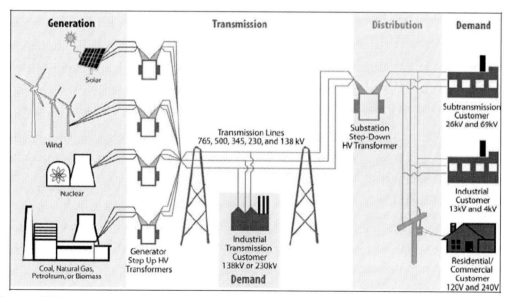

Source: CRS.

Figure A-1. Elements of the Electric Power System.

CAPACITY AND ENERGY

Capacity is the potential instantaneous output of a generating or storage unit, measured in watts. Energy is the actual amount of electricity generated by a power plant or released by a storage device during a time period, measured in watt-hours. The units are usually expressed in thousands (kilowatts and kilowatt-hours) or millions (megawatts and megawatt-hours). For example, the maximum amount of power a 1,000 megawatt (MW) power plant can generate in a year is 8.76 million megawatt-hours (MWh), calculated as: 1,000 MW x 8,760 hours in a year = 8.76 million MWh.

Capacity Factor

Capacity factor is a standard measure of how intensively a power plant is utilized. It is the ratio of how much electricity a power plant produced over a period of time, typically a year, compared to how much electricity the plant could have produced if it operated continuously at full output. For example, as shown in the prior paragraph, the maximum possible output of a 1,000 MW power plant in one year is 8.76 million MWh. Assume that during a year the plant actually produced only 7.0 million MWh. In this case the plant's capacity factor would be 7.0 million MWh ÷ 8.76 million MWh = 81%.

Generation and Load

The demand for electricity ("load") faced by an electric power system varies moment to moment with changes in business and residential activity and the weather. Load begins growing in the morning as people waken, peaks in the early afternoon, and bottoms-out in the late evening and early morning. **Figure A-2** shows an illustrative daily load curve.

The daily load shape dictates how electric power systems are operated. As shown in **Figure A-2**, there is a minimum demand for electricity that occurs throughout the day. This base level of demand is met with "baseload" generating units which have low variable operating costs.[40] Baseload units can also meet some of the demand above the base, and can reduce output when demand is unusually low. The units do this by "ramping" generation up and down to meet fluctuations in demand.

The greater part of the daily up and down swings in demand is met with "intermediate" units (also referred to as load-following or cycling units). These units can quickly change their output to match the change in demand (that is, they have a fast "ramp rate"). Load-following plants can also serve as "spinning reserve" units that are running but not putting power on the grid, and are immediately available to meet unanticipated increases in load or to back up other units that go off-line due to breakdowns.

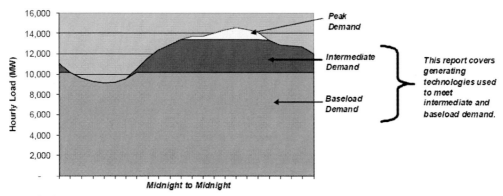

Source: CRS.

Figure A-2. Illustrative Daily Load Curve.

The highest daily loads are met with peaking units. These units are typically the most expensive to operate, but can quickly start up and shut down to meet brief peaks in demand. Peaking units also serve as spinning reserve and as "quick start" units able to go from shutdown to full load in minutes. A peaking unit typically operates for only a few hundred hours a year.

Economic Dispatch and Heat Rate

The generating units available to meet system load are "dispatched" (put on-line) in order of lowest variable cost. This is referred to as the "economic dispatch" of a power system's plants.

For a plant that uses combustible fuels (such as coal or natural gas) a key driver of variable costs is the efficiency with which the plant converts fuel to electricity, as measured by the plant's "heat rate." This is the fuel input in British Thermal Units (btus) needed to produce one kilowatt-hour of electricity output. A lower heat rate equates with greater efficiency and lower variable costs. Other things (most importantly, fuel and environmental compliance costs) being equal, the lower a plant's heat rate, the higher it will stand in the economic dispatch priority order. Heat rates are inapplicable to plants that do not use combustible fuels, such as nuclear and non-biomass renewable plants.

As an illustration of economic dispatch, consider a utility system with coal, nuclear, geothermal, natural gas combined cycle, and natural gas peaking units in its system:

1) Nuclear, coal, and geothermal baseload units, which are expensive to build but have low fuel costs and therefore low variable costs, will be the first units to be put online. Other than for planned and forced maintenance, these baseload generators will run throughout the year.
2) Combined cycle units, which are very efficient but use more expensive natural gas as a fuel, will meet intermediate load. These cycling plants will ramp up and down during the day, and will be turned on and off dozens of times a year.

3) Peaking plants, using combustion turbines,41 are relatively inefficient and burn natural gas. They run only as needed to meet the highest loads.42

An exception to this straightforward economic dispatch are "variable renewable" power plants—wind and solar—that do not fall neatly into the categories of baseload, intermediate, and peaking plants. Variable renewable generation is used as available to meet demand. Because these resources have very low variable costs they are ideally used to displace generation from gas-fired combined cycle plants and peaking units with higher variable costs. However, if wind or solar generation is available when demand is low (such as a weekend or, in the case of wind, in the evening), the renewable output could displace coal generation.

Power systems must meet all firm loads at all times, but variable renewable plants do not have firm levels of output because they depend on the weather. They are not firm resources because there is no guarantee that the plant can generate at a specific load level at a given point in time.[43] Variable renewable generation can be made firm by linking wind and solar plants to electricity storage, but with current technology, storage options are limited and expensive.

APPENDIX B. COMBINED CYCLE TECHNOLOGY

The combined cycle achieves a high level of efficiency by capturing waste heat that would otherwise be lost in the generating process. As shown in Figure B-1 for a combined cycle unit fueled by natural gas, the gas is fed into a combustion turbine which burns the fuel to power a generator. The exhaust from the combustion turbine is then directed to a specialized type of boiler (the heat recovery steam generator or HRSG) where the heat in the exhaust gases is used to produce steam, which in turn drives a second generator. In combined heat and power (CHP) applications, part of the steam is used to support an industrial process or to provide space heating, further increasing the total energy efficiency of the system.

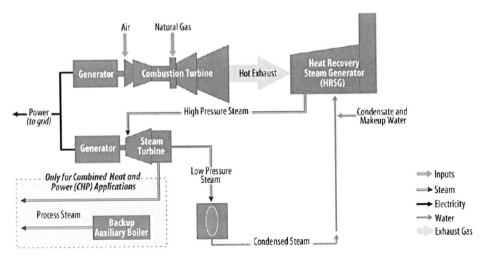

Source: CRS, based on a Calpine Corp. illustration.

Figure B-1. Schematic of a Combined Cycle Power Plant.

Combined cycles are built in different configurations, depending in part on the amount of capacity needed. **Figure B-1** illustrates a configuration in which one combustion turbine feeds one HRSG; this is referred to as "1x1" design. In higher capacity 2x1 or 3x1 designs, multiple combustion turbines feed a single HRSG. These options illustrate the modular (or "building block") nature of combined cycles, which facilitates rapid and flexible construction of new generating units to match changes in demand.

In the United States the predominant fuel used in combined cycle plants is natural gas. Combined cycles can also be designed to use fuel oil as a primary or backup fuel. Gasified coal can also be used as the fuel in an integrated gasification combined cycle (IGCC) plant. There are currently two prototype IGCC plant operating in the United States and a commercial-scale unit is under construction in Indiana.

End Notes

[1] Energy Information Administration (EIA), *Annual Energy Review 2008*, Tables 12.1 and 12.7b, http://www.eia.doe.gov/emeu/aer/envir.html.

[2] Written testimony of Dennis McConaghy, Executive Vice President, TransCanada Pipelines, Ltd., before the Senate

Energy and Natural Resources Committee hearing on *The Role of Natural Gas in Mitigating Climate Change*, October 28, 2009, p. 6, http://energy.senate.gov/public/index.cfm?FuseAction= Hearings.Hearing& Hearing_ID=788a1684- b2a2-f5bb-f574-81b9257ba5aa.

[3] Written testimony of Jack Fusco, President and CEO, Calpine Corp., before the Senate Energy and Natural Resources Committee hearing on *The Role of Natural Gas in Mitigating Climate Change*, October 28, 2009, p. 1, http://energy.senate.gov/public/index.cfm?FuseAction=Hearings.Hearing&Hearing_ID=788a1684-b2a2-f5bb-f574- 81b9257ba5aa.

[4] "Capacity" is a measure of the potential instantaneous electricity output from a power plant, usually measured in megawatts or kilowatts. "Generation" is the actual amount of electricity produced by the plant over a period of time, usually measured in megawatt-hours or kilowatt-hours. For additional information see Appendix A.

[5] These four sources accounted for 96% of electricity production in 2008, which is the typical combined share going back to the 1980s. All other sources, such as wind, petroleum, and biomass, account for the remaining 4%.

[6] By extracting the maximum energy from fuel combustion, modern combined cycles can reportedly achieve heat rates in the range of 6,752 to 6,333 btus per kwh. This compares to 9,200 to 8,740 btus per kwh for steam electric coal technology. (EIA, *Assumptions to the Annual Energy Outlook 2009*, Table 8.2, http://www.eia.doe.gov/oiaf/aeo/assumption/index.html.) This efficiency advantage can make combined cycles very economical to operate.

[7] For more information on power plant cost and construction issues, see CRS Report RL34746, *Power Plants: Characteristics and Costs*, by Stan Mark Kaplan.

[8] Rebecca Smith, "Utilities Question Natural-Gas Forecasting," *The Wall Street Journal*, December 27, 2004.

[9] Natural gas emits 117.08 pounds CO2 per MMBtu burned. The comparable numbers for subbituminous, bituminous, and lignite coal are, respectively, 212.7, 205.3, and 215.4 pounds of CO2 per MMBtu burned. EIA, *Electric Power Annual 2007*, Table A3, http://www.eia.doe.gov/cneaf/electricity/epa/epa_sum.html.

[10] Other environmental advantages of combined cycle plants include minimal or zero emissions of sulfur dioxides and mercury; no solid wastes, such as ash and scrubber sludge; no coal piles with attendant fugitive dust and runoff problems; and the fuel is delivered by pipeline rather than railroad or truck. A combustion turbine burning natural gas will emit more nitrogen oxides (NOx) per MMBtu of fuel consumed than a coal-fired boiler, and depending on the location of a gas-fired plant it may have to install a low NOx combustion system and a selective catalytic reduction system to capture NOx emissions.

[11] The study group of combined cycle plants consists of plants with the following characteristics: minimum net summer capacity of 100 MW; the plant operated at some point in time during 2007 and was in operational condition at the end of 2007; the plant's primary fuel was natural gas; and the plant's primary purpose was to sell power to the public. (This last criterion excludes industrial and commercial cogenerators who operate a power plant primarily to provide electricity and steam to a single business establishment.) A total of 314

combined cycle plants with total capacity of 170,627 MW met these criteria. The study group of coal plants had the same criteria except that the capacity floor was 250 MW and the primary fuel had to be coal or waste coal. A total of 298 coal plants with total capacity of 284,646 MW met these criteria. CRS identified the plants and extracted the data from the EIA-860 and EIA-906/920 databases for 2007 (http://www.eia.doe.gov/cneaf/electricity/page/data.html). The 2007 generation from the plants in the combined cycle study group (630.4 million MWh) accounted for 98% of all gas-fired combined cycle generation in the electric power sector in 2007. Similarly, the generation from the coal plants in the study group (1,870.6 million MWh) accounted for 95% of all coal-fired generation in the electric power sector.

[12] Capacity factor is a measure of the actual utilization of a power plant compared to its hypothetical maximum utilization. For additional information see Appendix A.

[13] For information on the study group of coal plants see footnote 11. Capacity factors were calculated using the EIA906/920 generation and EIA-860 generating capacity databases (http://www.eia.doe.gov/cneaf/electricity/page/ data.html).

[14] Line loss is the loss of electrical energy due to the resistance of the length of wire in a circuit. Much of the loss is thermal in nature. (This definition is a composite created from the glossaries at http://www.eia.doe.gov/glossary/ glossary_l.htm and http://www.ewh.ieee.org/sb/srisairamec/glossary/k-lglos.htm.)

[15] Most of the transmission grid uses alternating current (AC) technology which is prone to line losses. By using transmission lines with higher kilovolt (kV) ratings, more power can be transported long distances with fewer losses. The highest capacity AC lines currently in use in the United States have a rating of 765 kV, but according to DOE these lines make up less than 2% of the grid. An alternative, direct current (DC) technology, can move large amounts of power long distances with minimal losses. However, DC lines are in only limited use (about 2% of the grid) because they are more difficult and expensive to integrate into the grid than AC lines. Proposals have been made to upgrade the AC network by building more 765 kV lines and lines using even high capacity AC technology (referred to as ultra high voltage transmission), and to build more DC lines. These proposals are generally focused on moving renewable power long distances. For example, see American Electric Power, *Interstate Transmission Vision for Wind Integration*, undated, http://www.aep.com/about/i765project/docs/WindTransmissionVisionWhitePaper.pdf, and the Joint Coordinated System Plan proposal at http://www.jcspstudy.org/. (The transmission line statistics cited in this footnote are from DOE, *National Transmission Grid Study*, May 2002, p. 3, http://www.ferc.gov/industries/electric/indus-act/transmission-grid.pdf.)

[16] Ventyx Corp., *Major Transmission Constraints in PJM*, 2007, http://www1.ventyx.com/pdf/wp07-transmission-constraints.pdf.

[17] Western Electric Coordinating Council, *2008 Annual Report of the Transmission Expansion Planning Policy Committee, Executive Summary*, March 31, 2009, p.9, http://www.wecc.biz/committees/BOD/TEPPC/Shared%20Documents/TEPPC%20Annual%20Reports/2008/CoverLetter_Exec_Summary_Final_.pdf.

[18] Connecticut General Assembly, Office of Legislative Research, *Factors Behind Connecticut's High Electric Rates*, August 5, 2008, No. 2008-R-0452, http://www.cga.ct.gov/2008/rpt/2008-R-0452.htm.

[19] Bernard Lesieutre and Joseph Eto, *Electricity Transmission Congestion Costs: A Review of Recent Reports*, Lawrence Berkeley National Laboratory, p. 2, http://certs.lbl.gov/pdf/54049.pdf, and U.S. Department of Energy, *National Transmission Grid Study*, May 2002, pp. 16–18, http://www.pi.energy.gov/documents/TransmissionGrid.pdf.

[20] U.S. DOE, *National Electric Transmission Congestion Study*, August 2008, p. 28, http://www.pi.energy.gov/documents/TransmissionGrid.pdf. Emphasis in the original not shown.

[21] The study group of combined cycle plants includes 314 larger plants that accounted for 98% of combined cycle generation in the electric power sector in 2007. The study group of coal plants includes 298 larger plants that accounted for 95% of coal-fired generation in the electric power sector in 2007. For additional information on the characteristics of the study groups see footnote 11.

[22] Actual generation in 2007 is from the EIA-906/920 database. Capacity factors were computed using this generation data and each plant's capacity as reported in the EIA-860 database. An NGCC plant was assumed to have surplus generation if its annual capacity factor in 2007 was less than 85%; that is, the hypothetical surplus generation available to displace coal was the difference between the NGCC plant's actual generation in 2007 and the electricity it could have produced at an 85% utilization rate. A few NGCC plants had capacity factors of 85% or greater in 2007 and were therefore assumed to have no surplus generation available for coal displacement. The EIA databases are available at http://www.eia.doe.gov/cneaf/electricity/page/data.html.

[23] CO2 emissions were estimated for each coal plant based on the type and volume of coal consumed. Fuel consumption in MMBtus was taken from the EIA-906/920 database and used to calculate CO2 emissions using the emission factors in EIA, *Electric Power Annual 2007*, Table A3, http://www.eia.doe.gov/cneaf/electricity/epa/epa_sum.html. The same data sources were used to calculate CO2 emissions for combined cycles.

[24] The maps only show coal plants assumed to have had generation displaced, and the existing NGCC plants responsible for the displacement. If a coal plant had an NGCC plant with the ten or 25 mile radius, but the NGCC plant was assumed to be unavailable to displace coal (for example, because it had a capacity factor in 2007 of 85% or higher) no coal is assumed to have been displaced and the coal plant is not shown on the map.

[25] According to U.S. EPA data, the gross output of the NGCC units at Plant Barry was 288,726 MWh in November 2007. In comparison, the highest output was 532,040 MWh in August. The data was downloaded from the EPA website at http://camddataandmaps.epa.gov/gdm/index.cfm? Fuseaction =emissions.prepackaged_select.

[26] The NGCC units at Plant Barry have, according to the Platts Powermap database, a nominal total net winter capacity of 1,090 MW. However, the maximum output achievable at any point in time will vary with the ambient air temperature, which affects the density of the air flow into the combustion turbine units of a NGCC.

[27] For additional information see CRS Report R40894, *Unconventional Gas Shales: Development, Technology, and Policy Issues*, coordinated by Anthony Andrews; and FERC, *State of the Markets Report 2008*, August 2009, Section 2, http://www.ferc.gov/market-oversight/st-mkt-ovr/2008-som-final.pdf.

[28] EIA, *Short-Term Energy Outlook*, December 2009, pp. 4 – 6, http://www.eia.doe.gov/emeu/steo/pub/dec09.pdf.

[29] FERC, *Winter 2009/2010 Energy Market Assessment*, November 19, 2009, p. 3, http://www.ferc.gov/EventCalendar/Files/20091119102759-A-3-final.pdf. For additional information on the findings of the Potential Gas Committee, see the press release at http://www.energyindepth.org/wp-content/uploads/2009/03/potential-gas-committee-reportsunprecedented-increase-in.pdf.

[30] EIA, *Annual Energy Outlook 2010 Early Release Overview*, pp. 3, 4, 12, http://www.eia.doe.gov/oiaf/aeo/pdf/overview.pdf.

[31] EIA, *Annual Energy Outlook Retrospective Review: Evaluation of Projections in Past Editions (1982-2008)*, September 2008, p. 2, http://www.eia.doe.gov/oiaf/analysispaper/retrospective/pdf/0640%282008%29.pdf.

[32] Rebecca Smith, "Utilities Question Natural-Gas Forecasting," *The Wall Street Journal*, December 27, 2004.

[33] Statement for the Record of Edward Stones, Director of Risk Management, Dow Chemical Co., before the Senate Energy and Natural Resources Committee hearing on *The Role of Natural Gas in Mitigating Climate Change*, October 28, 2009, p. 4, http://energy.senate.gov/public/_files/StonesTestimony102809.pdf.

[34] Ibid., p. 3.

[35] EIA's Annual Energy Outlook 2010 backup spreadsheet for Table 13, located at http://www.eia.doe.gov/oiaf/aeo/ aeoref_tab.html. Values were converted from cubic feet to Btus using a conversion factor of 1.028. This is the Reference Case forecast, which assumes no changes to current law or regulations.

[36] The interstate portion of the system consists of 217,000 miles of pipeline with a capacity of 183 BCF per day. The interstate portion consists of 76,000 miles of pipeline with a capacity of 32 BCF per day. EIA, *Expansion of the U.S. Natural Gas Pipeline Network: Additions in 2008 and Projects through 2011*, September 2009, p. 3, http://www.eia.doe.gov/pub/oil_gas/natural_gas/feature_articles/2009/pipelinenetwork/pipeline network.pdf.

[37] As shown in Table 3, the maximum potential increase in existing NGCC generation to displace coal is 640,128,780 MWh. This number assumes an annual average capacity factor of 85%, but on a given day the existing NGCC plants could be running at full load to displace coal, which is 640,128,780 MWh ÷ 0.85 ÷ 365 days = 2,063,268 MWh per day. The average heat rate for combined cycles in the study group is 7.4596 MMBtus per MWh and the conversion factor from MMBtus of thousands of cubic feet is 1.028, so the daily gas demand can be calculated as 2,063,268 MWh x 7.4596 MMBtus per MWh ÷ 1.028 ÷ 1,000,000 = 15 BCF per day.

[38] EIA, *Expansion of the U.S. Natural Gas Pipeline Network: Additions in 2008 and Projects through 2011*, September 2009, Table 2, http://www.eia.doe.gov/pub/oil_gas/natural_gas/ feature_articles/ 2009/pipeline network/pipelinenetwork.pdf.

[39] Many coal plants use natural gas as a startup fuel and for flame stabilseation during normal operations. However, this is different from running the plant primarily or largely on natural gas. In addition to the engineering issues, even if a coal plant currently uses natural gas as a startup fuel, its existing natural gas pipeline connection may not have sufficient capacity to provide enough gas for full load (or even large partial load) operation on gas. There are examples of coal units switching to natural gas for environmental reasons

("Marketwatch: Public Service Electric & Gas of New Jersey," *Platts Coal Week*, June 22, 1992; "Ill. Power to Shift Vermilion to Gas; Phase I Decision Kills Coal Solicitation," *Platts Coal Week*, October 10, 1994; "PEPCO Mulls NOx Ozone Season's Effect on Coal, Gas, Oil Use," *Platts Coal Week*, October 25, 1999.

[40] Variable costs are costs that vary directly with changes in output. For fossil fuel units the most important variable cost is fuel. Solar and wind plants have minimal or no variable costs, and nuclear plants have low variable costs.

[41] A combustion turbine is an adaption of jet engine technology to electric power generation. A combustion turbine can either be used stand-alone as a peaking unit, or as part of a more complex combined cycle plant used to meet intermediate and baseload demand.

[42] This alignment of generating technologies is for new construction using current technology. The existing mix of generating units in the United States contains many exceptions to this alignment of load to types of generating plants, due to changes in technology and economics. For instance, there are natural gas and oil-fired units built decades ago as baseload stations that now operate as cycling or peaking plants because high fuel prices and poor efficiency has made them economically marginal. Some of these older plants were built close to load centers and are now used as reliability must-run (RMR) generators that under certain circumstances must be operated, regardless of cost, to maintain the stability of the transmission grid.

[43] Hydroelectric generation is a special case. Hydro generation is very low cost and is firm, dispatchable capacity to the degree there is water in the dam's reservoir. However, operators have to consider not only how much water is currently available, but how much may be available in upcoming months, and competing demands for the water, such as drinking water supply, irrigation, and recreation. These factors can make hydro dispatch decisions very complex. In general hydro is used to meet load during high demand hours, when it can displace expensive peaking and cycling units, but if hydro is abundant it can also displace baseload coal plants.

Natural Gas: Outlooks and Opportunities
Editor: Lucas N. Montauban

ISBN: 978-1-61324-132-5
© 2011 Nova Science Publishers, Inc.

Chapter 3

FEDERAL OIL AND GAS LEASES: OPPORTUNITIES EXIST TO CAPTURE VENTED AND FLARED NATURAL GAS, WHICH WOULD INCREASE ROYALTY PAYMENTS AND REDUCE GREENHOUSE GASES[*]

United States Government Accountability Office
Report to Congressional Requesters

WHY GAO DID THIS STUDY

The Department of the Interior (Interior) leases public lands for oil and natural gas development, which generated about $9 billion in royalties in 2009. Some gas produced on these leases cannot be easily captured and is released (vented) directly to the atmosphere or is burned (flared). This vented and flared gas represents potential lost royalties for Interior and contributes to greenhouse gas emissions.

GAO was asked to (1) examine available estimates of the vented and flared natural gas on federal leases, (2) estimate the potential to capture additional gas with available technologies and associated potential increases in royalty payments and decreases in greenhouse gas emissions, and (3) assess the federal role in reducing venting and flaring. In addressing these objectives, GAO analyzed data from Interior, the Environmental Protection Agency (EPA), and others and interviewed agency and industry officials.

WHAT GAO RECOMMENDS

To reduce lost gas, increase royalties, and reduce greenhouse gas emissions, GAO recommends that Interior improve its venting and flaring data and address limitations in its regulations and guidance. Interior generally concurred with these recommendations.

---

[*] This is an edited, reformatted and augmented version of a United States Government Accountability Office's publication, dated October 2010.

WHAT GAO FOUND

Estimates of vented and flared natural gas for federal leases vary considerably, and GAO found that data collected by Interior to track venting and flaring on federal leases likely underestimate venting and flaring because they do not account for all sources of lost gas. For onshore federal leases, operators reported to Interior that about 0.13 percent of produced gas was vented or flared. Estimates from EPA and the Western Regional Air Partnership (WRAP) showed volumes as high as 30 times higher. Similarly, for offshore federal leases, operators reported that 0.5 percent of the natural gas produced was vented and flared, while data from an Interior offshore air quality study showed that volume to be about 1.4 percent, and estimates from EPA showed it to be about 2.3 percent. GAO found that the volumes operators reported to Interior do not fully account for some ongoing losses such as the emissions from gas dehydration equipment or from thousands of valves—key sources in the EPA, WRAP, and Interior offshore air quality studies.

Source: EPA.

Vented Gas from Oil Storage Tank Visible through Infrared Camera.

Data from EPA, supported by information obtained from technology vendors and GAO analysis, suggest that around 40 percent of natural gas estimated to be vented and flared on onshore federal leases could be economically captured with currently available control technologies. According to GAO analysis, such reductions could increase federal royalty payments by about $23 million annually and reduce greenhouse gas emissions by an amount equivalent to about 16.5 million metric tons of CO_2—the annual emissions equivalent of 3.1 million cars. Venting and flaring reductions are also possible offshore, but data were not available for GAO to develop a complete estimate.

As part of its oversight responsibilities, Interior is charged with minimizing vented and flared gas on federal leases. To minimize lost gas, Interior has issued regulations and guidance that limit venting and flaring during routine procedures. However, Interior's oversight efforts to minimize these losses have several limitations, including that its regulations and guidance do not address some significant sources of lost gas, despite available control technologies to potentially reduce them. Although EPA does not have a role in managing federal leases, it has voluntarily collaborated with the oil and gas industry through its Natural Gas STAR program, which encourages oil and gas producers to use gas saving technology, and through which operators reported venting reductions totaling about 0.4 percent of natural gas production in 2008.

ABBREVIATIONS

Bcf	billion cubic feet
BLM	Bureau of Land Management
BOEMRE	Bureau of Ocean Energy Management, Regulation and Enforcement
EIA	Energy Information Administration
EPA	Environmental Protection Agency
GOADS	Gulfwide Offshore Activities Data System
Interior	Department of the Interior
MRM	Minerals Revenue Management
NTL	Notice to Lessees and Operators
OGOR	Oil and Gas Operations Report
WRAP	Western Regional Air Partnership
VOC	volatile organic compound

October 29, 2010
The Honorable Darrell Issa
Ranking Member
Committee on Oversight
and Government Reform
House of Representatives

The Honorable Nick J. Rahall, II
Chairman
Committee on Natural Resources
House of Representatives

Production of oil and natural gas on federal lands and waters is an important part of the nation's energy portfolio and a significant source of revenue for the federal government. The Department of the Interior (Interior) manages lands that account for nearly a quarter of domestic oil and gas production. In fiscal year 2009, companies that leased these lands paid about $6 billion in royalties to the federal government on the sale of oil and gas produced offshore in federal waters, and about $3 billion for production on federal lands, making revenues from federal oil and gas one of the largest nontax sources of federal government funds. Interior's Bureau of Land Management (BLM) is responsible for managing leases onshore, and its Bureau of Ocean Energy Management, Regulation and Enforcement (BOEMRE) is responsible for leases offshore.[1]

While most of the natural gas produced on leased federal lands and waters is sold, some is lost during production for various reasons, including leaks and releases for ongoing operational or safety procedures. This natural gas is either released directly into the atmosphere (vented) or burned (flared).[2] The venting and flaring of natural gas is the potential loss of a valuable resource and, on leased federal lands or waters, the loss of federal royalty payments. In addition, venting releases methane, and flaring emits carbon dioxide (CO_2), both greenhouse gases that contribute to global climate change. Methane is a particular concern since it is a more potent greenhouse gas than is CO_2.[3]

In 2004, we reported that Interior, the department charged with managing federal oil and gas leases and regulating venting and flaring, collected and reported information on the extent of venting and flaring on leased federal lands and waters.[4] We made two recommendations to Interior to improve the measurement of vented and flared gas and to reduce its impact, which the department implemented. Since that time, the Environmental Protection Agency (EPA) and the oil and gas industry identified sources of vented and flared gas that were releasing substantially more gas than previously thought possible, suggesting that the expanded use of available technologies could help capture additional gas. This report responds to your request that we review the extent of venting and flaring of natural gas on federal leases. Our objectives were to (1) examine available estimates of vented and flared natural gas on federal leases; (2) estimate the potential to capture additional vented and flared natural gas with available technologies and the associated potential increases in royalty payments and reductions in greenhouse gas emissions and; (3) assess the federal role in reducing venting and flaring of natural gas.

To examine estimates of the volumes of vented and flared natural gas on federal leases, we analyzed data on venting and flaring that oil and gas producers submit to Interior's Minerals Revenue Management (MRM) program, which is responsible for collecting revenue from federal leases. MRM uses these data from its Oil and Gas Operations Report (OGOR) data system to account for monthly oil and gas production onshore and offshore. Separate regulations and guidance from BLM and BOEMRE guide operators in reporting to OGOR, and these data are the primary information source these agencies use to monitor overall venting and flaring. We also analyzed data from BOEMRE's Gulfwide Offshore Activity Data System (GOADS), which BOEMRE collects and publishes in a study every 3 years.[5] BOEMRE uses its GOADS studies to estimate the impacts of offshore oil and gas exploration, development, and production on onshore air quality in the Gulf of Mexico region, which made up about 98 percent of federal offshore gas production in 2008. BOEMRE also uses GOADS as part of an impact analysis required by the National Environmental Policy Act. In addition, we analyzed EPA estimates of vented and flared gas onshore and offshore.[6] We also analyzed data on vented and flared natural gas that the Western Regional Air Partnership (WRAP),[7] in conjunction with the Independent Petroleum Association of Mountain States,[8] collected from the oil and gas industry to measure air quality in a number of large production basins in the mountain west.[9] We had consultants from the Environ International Corporation, the firm that collected and analyzed the air quality data for WRAP, reconfigure these data to provide information on venting and flaring volumes on federal leases for a number of these onshore basins. Our sources of venting and flaring data were from 2006 to 2008, and we examined only the portions of these data related to federal leases in order to ensure comparability between them. We assessed the reliability of the data we used by analyzing the methods used to construct them and found them sufficiently reliable for the purposes of this report.

To estimate the potential federal royalty increases and greenhouse gas reductions resulting from capturing additional vented and flared gas with available technologies, we met and spoke with officials from the oil and gas industry and to vendors of products designed to reduce vented and flared gas about how and under what conditions these technologies could reduce venting and flaring. We used analyses and data from EPA and WRAP to estimate potential reductions in volumes of vented and flared natural gas on federal leases, then converted these volumes into potential federal royalty increases and greenhouse gas reductions using methane to carbon dioxide equivalent conversion factors, average natural gas prices, and royalty rates from Interior. To assess the federal role in reducing vented and flared gas, we interviewed officials from BLM and BOEMRE, including officials from field offices that manage oil and gas leases in large onshore and offshore production basins; EPA; the Department of Energy; state agencies; and the oil and gas industry. We also reviewed BLM and BOEMRE regulations and other documentation, other studies related to federal management and oversight of the oil and gas industry, as well as a prior GAO report that described limitations in the systems Interior has in place to track oil and gas production on federal leases.[10] See appendix I for more detailed information on our scope and methodology.

We conducted this performance audit from July 2009 to October 2010 in accordance with generally accepted government auditing standards. Those standards require that we plan and perform the audit to obtain sufficient, appropriate evidence to provide a reasonable basis for our findings and conclusions based on our audit objectives. We believe that the evidence obtained provides a reasonable basis for our findings and conclusions based on our audit objectives.

BACKGROUND

The Mineral Leasing Act of 1920 charges Interior with overseeing oil and gas leasing on federal lands and private lands where the federal government has retained mineral rights covering about 700 million onshore acres.[11] Offshore, the Outer Continental Shelf Lands Act,[12] as amended, gives Interior the responsibility for leasing and managing approximately 1.76 billion acres. BLM and BOEMRE are responsible for issuing permits for oil and gas drilling; establishing guidelines for measuring oil and gas production; conducting production inspections; and generally providing oversight for ensuring that oil and gas companies comply with applicable laws, regulations, and department policies. This oversight includes the authority to ensure that firms produce oil and gas in a manner that minimizes any waste of these resources. Together, BLM and BOEMRE are responsible for oversight of oil and gas operations on more than 28,000 producible leases.

Interior's MRM program, which is managed under BOEMRE, is charged with ensuring that the federal government receives royalties from the operators that produce oil and gas from both onshore and offshore federal leases. MRM is responsible for collecting royalties on all of the oil and gas produced, with some allowances for gas lost during production. Companies pay royalties to MRM based on a percentage of the cash value of the oil and gas produced and sold. Currently, royalty rates for onshore leases are generally 12.5 percent, while rates for offshore leases range from 12.5 percent to 18.75 percent.

The production of oil and gas on these federal leases involves several stages, including the initial drilling of the well; clearing out liquid and mud from the wellbore; production of oil and gas from the well; separation of oil, gas, and other liquids; transfer of oil and gas to storage tanks; and distribution to central processing facilities. Throughout this process, operators typically vent or flare some natural gas, often intermittently in response to maintenance needs or equipment failures. This intermittent venting may take place when operators purge water or hydrocarbon liquids that collect in well bores (liquid unloading) to maintain proper well function or when they expel liquids and mud with pressurized natural gas after drilling during the well completion process. BLM and BOEMRE permit operators of wells to release routine amounts of gas during the course of production without notifying them or incurring royalties on this gas.[13] In addition, production equipment often emits gas to maintain proper internal pressure, or in some cases, the release of pressurized gas itself is the power source for the equipment, particularly in remote areas that are not linked to an electrical grid. This "operational" venting may include the continuous releases of gas from pneumatic devices—valves that control gas flows, levels, temperatures, and pressures in the equipment and rely on pressurized gas for operation—as well as leaks, or "fugitive" emissions.[14] It also includes natural gas that vaporizes from oil or condensate storage tanks or during the normal operation of natural gas dehydration equipment.[15] Until recently, the industry considered these operational losses to be small, but recent infrared camera technology has shed new light on these sources of vented gas, particularly from condensate storage tanks.[16] According to oil and gas industry representatives, the cameras helped reveal that losses from storage tanks and fugitive emissions were much higher than they originally thought (link to video).[17] In addition, recent calculations from EPA suggest that emissions from completions and liquid unloading make larger contributions to lost gas than previously thought possible. Operators can use a number of techniques to estimate emissions based on gas and oil characteristics and well operating conditions, such as temperature and pressure, without taking direct measurements of escaping gas.

While venting and flaring of natural gas is often a necessary part of production, the lost gas has both economic and environmental implications. On federal oil and gas leases, natural gas that is vented or flared during production, instead of captured for sale, represents a loss of royalty revenue for the federal government.[18] Venting and flaring natural gas also adds to greenhouse gases in the atmosphere. In general, flaring emits CO_2, while venting releases methane, both of which the scientific community agrees are contributing to global warming. Methane is considered particularly harmful in this respect, as it is roughly 25 times more potent by weight than CO_2 in its ability to warm the atmosphere over a 100-year period and almost 72 times more potent over a 20-year period, according to the Intergovernmental Panel on Climate Change.[19] Other hydrocarbons and compounds in vented and flared gas can also harm air quality by increasing ground-level ozone levels and contributing to regional haze. Volatile organic compounds, present in vented gas, are contributors to elevated ozone and haze, and ozone is a known carcinogen, according to EPA analysis.[20] In some areas in the western United States, the oil and gas industry is a major source of volatile organic compounds. According to EPA, in many western states, including in many rural areas where there is substantial oil and gas production and limited population, there have been increases in ozone levels, often exceeding federal air quality limits.[21] Interior is required to conduct environmental impact assessments in advance of oil and gas leasing and generally works with state environmental and air quality agencies to ensure that oil and gas producers will comply with environmental laws such as the Clean Air Act or Clean Water Act and the related implementing regulations. However, the state agencies may be charged with maintaining the standards established by the federal government in law and regulation, and often have primary responsibility in this regard.[22]

While much of the natural gas that is vented and flared is considered to be unavoidably lost, certain technologies and practices can be applied throughout the production process to capture some of this gas according to the oil and gas industry and EPA. The technologies' technical and economic feasibility varies and sometimes depends on the characteristics of the production site. For example, some technologies require a substantial amount of electricity, which may be less feasible for remote production sites that are not on the electrical grid. However, certain technologies are generally considered technically and economically feasible at particular production stages, including the following:[23]

- *Drilling:* Using "reduced emission" completion equipment when cleaning out a well before production, which separates mud and debris to capture gas or condensate that might otherwise be vented or flared.
- *Production:* Installing a plunger lift system to facilitate liquid unloading. Plunger-lift systems drop a plunger to the bottom of the well, and when the built-up gas pressure pushes the plunger to the surface, liquids come with it. Most of the accompanying gas goes into the gas line rather than being vented. Computerized timers adjust when the plunger is dropped according to the rate at which liquid collects in the well, further decreasing venting.
- *Storage:* Installing vapor recovery units that capture gas vapor from oil or condensate storage tanks and send it into the pipeline.
- *Dehydration:* Optimizing the circulation rate of the glycol and adding a flash tank separator that reduces the amount of gas that is vented into the atmosphere.[24]
- *Pneumatic devices:* Replacing pneumatic devices at all stages of production that release, or "bleed," gas at a high rate (high-bleed pneumatics) with devices that bleed gas at a lower rate (low-bleed pneumatics).

In 2004, we reported that information on the extent to which venting and flaring occurs was limited.[25] Although BLM and BOEMRE require operators to report data on venting and flaring on a monthly basis, our 2004 report found that these data did not distinguish between gas that is vented and gas that is flared, making it difficult to accurately identify the extent to which each occurs. In implementing our recommendations for offshore operators, BOEMRE now requires operators to report venting and flaring separately and to install meters to measure this gas on larger platforms.[26] The Energy Information Administration (EIA)[27] also collects data from oil and gas producing states on venting and flaring, but our 2004 work found that EIA did not consider these state-reported data to be consistent and, according to discussions with EIA officials, these data have not improved.

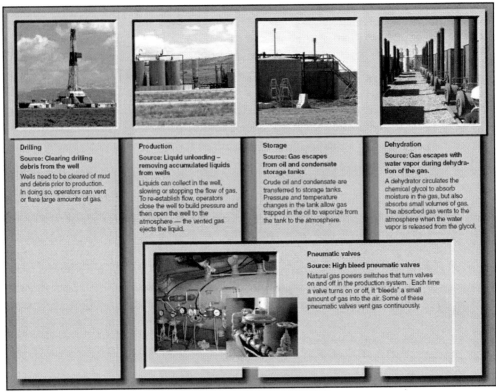

source: GAO; photos from Bureau of Land Management; storage photo from Wyoming Department of Environment Quality.

Figure 1. Illustrative Example of Onshore Production and Associated Sources of Vented and Flared Gas.

AVAILABLE ESTIMATES OF VENTED AND FLARED NATURAL GAS VARY, BUT VOLUMES REPORTED TO OGOR ARE LIKELY UNDERESTIMATED BECAUSE THEY DO NOT INCLUDE SOME SOURCES

Available estimates of vented and flared natural gas on federal leases vary considerably, and we found that estimates based on data from MRM's OGOR data system likely underestimate these volumes because they include fewer sources of emissions than other estimates, including EPA's and WRAP's. For onshore federal leases, operators reported to OGOR that about 0.13 percent of the natural gas produced was vented and flared, while EPA estimates showed the volume to be about 4.2 percent, and estimates based on WRAP data showed it to be as high as 5 percent. Similarly, for offshore federal leases, operators reported to OGOR that 0.5 percent of the natural gas produced was vented and flared, while data in BOEMRE's GOADS system—a database that focuses on the impacts of offshore oil and gas exploration, development, and production on air quality in the Gulf of Mexico region— showed that volume to be about 1.4 percent, and estimates from EPA showed it to be about 2.3 percent.[28]

Onshore Leases

Onshore leases showed the largest variation between OGOR data and others' estimates of natural gas venting and flaring. Operators reported to MRM's OGOR system that about 0.13 percent of the natural gas produced on onshore federal leases was vented or flared each year between 2006 and 2008.[29] BLM uses guidance from 1980, which sets limits on the amount of natural gas that may be vented and flared on onshore leases, requires operators to report vented and flared gas to OGOR, and in some cases to seek permission before releasing gas.[30] Although the guidance states that onshore operators must report all volumes of lost gas to OGOR, it does not enumerate the sources that should be reported or specify how they should be estimated.[31] Staff from BLM told us that the reported volumes were from intermittent events like completions, liquid unloading, or necessary releases after equipment failures; however, operators did not report operational sources such as venting from oil storage tanks, pneumatic valves, or glycol dehydrators. In general, BLM staff said that they thought that vented and flared gas did not represent a significant loss of gas on federal leases. In addition, we found a lack of consistency across BLM field offices regarding their understanding of which intermittent volumes of lost gas should to be reported to OGOR. For example, staff from some of the offices said that they thought that intermittent vented and flared gas was not to be reported if operators had advance permission or where volumes were under BLM's permissible limits, while others said that they thought that operators still needed to report this gas. Our discussions with operators reflected this lack of consistency from BLM field office staff. Operators we spoke with said that they generally did not report operational sources, and in some cases did not report intermittent sources as long as they were under BLM's permissible limits for venting and flaring.[32]

In contrast, EPA's estimate of venting and flaring was approximately 4.2 percent of gas production on onshore federal leases for the same period and consistently included both intermittent and operational sources. EPA estimated these emissions using data on average nationwide oil and gas production equipment and their associated emissions (see table 1).[33] As noted earlier, venting from operational sources had not previously been seen as a significant contributor to lost gas. With these additional sources, EPA's estimates are around 30 times higher than the volumes operators reported to OGOR. According to EPA's estimates, the amount of natural gas vented and flared on onshore leases totaled around 126 billion cubic feet (Bcf) of gas in 2008. This amount is roughly equivalent to the natural gas needed to heat about 1.7 million homes during a year, according to our calculations. See figure 2 for a comparison between EPA's estimated gas emissions and the volumes reported to OGOR as a percentage of gas production on federal onshore leases.

Similarly, analysis of WRAP data for five production basins in the mountain west in 2006 indicated as much as 5 percent of the total natural gas produced on federal leases was vented and flared. WRAP based its estimates, in part, on a survey of the types of equipment operators were using,[34] and provided a detailed list of sources to be reported. WRAP's data included similar sources as EPA's data, as well as estimates of emissions from fugitive sources like leaking seals and valves. Although estimates based on WRAP data varied from basin to basin—between 0.3 and 5 percent—they were consistently much higher than the volumes operators reported to OGOR. The average vented and flared gas as a percentage of production was 2.2 percent across the five basins.[35] See table 2 for a list of the key sources in one of the five basins.

Table 1. EPA's Estimates of Vented and Flared Natural Gas and Sources for Onshore Federal Leases

Sources (2008)	Volume (Bcf)
Flared (variety of sources)	28
Pneumatic devices	16
Gas well liquid unloading	17
Well completions	30
Oil and condensate storage tanks	18
Glycol dehydrators	7
Other	10
Total	126

Source: GAO analysis of EPA data.

Note: Volatile organic compounds accounted for 26 Bcf of these emissions and were mostly from storage tanks and dehydrators according to EPA.

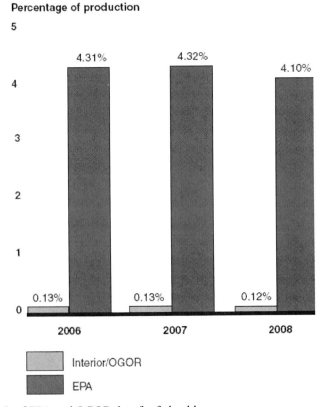

Source: GAO analysis of EPA and OGOR data for federal leases.

Figure 2. Comparison of OGOR Reported Volumes to EPA's Estimates of Vented and Flared Natural Gas for Onshore Federal Leases.

Table 2. Estimates of Vented and Flared Natural Gas based on 2006 WRAP Data for Federal Leases in the Piceance Basin (Colo.)

Sources from Piceance Basin	Volume (Bcf)
Well completions	2.4
Pneumatic devices	0.5
Gas well liquid unloading	0.4
Fugitive emissions	0.1
Condensate storage tanks	0.1
Other sources	0.4
Total	3.8

Source: Environ Corp. analysis of 2006 WRAP data for federal leases.

Note: Flared gas is included throughout several source categories, including completions and storage tanks. We chose to present data from the Piceance basin because it was representative of the key sources common to the other basins. Volume figures in table do not sum to 3.8 Bcf due to rounding.

In figure 3, which compares estimates based on WRAP data with the volumes operators reported to OGOR for 2006, for the Uinta basin, the WRAP estimate was about 20 times higher than the volumes reported to OGOR, and for two other basins (i.e., Denver-Julesburg and N. San Juan) no volumes of vented and flared gas were reported to OGOR.[36]

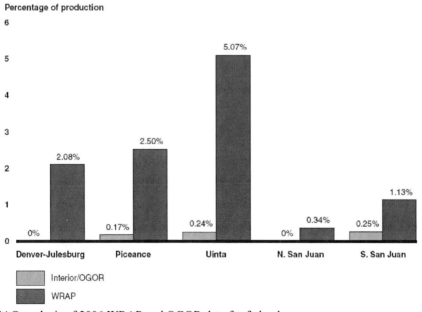

Source: GAO analysis of 2006 WRAP and OGOR data for federal

Note: The Denver-Julesburg, North San Juan, and Piceance basins are in Colorado, the Uinta basin is in Utah, and the South San Juan is in New Mexico. See appendix I for a map of the basins.

Figure 3. Comparison of 2006 OGOR-Reported Volumes to Estimates Based on 2006 WRAP Data of Vented and Flared Natural Gas for Onshore Federal Leases in Five Basins.

Offshore Leases

Offshore leases showed less variation between OGOR data and others' estimates of natural gas venting and flaring than onshore leases, but the volumes that operators reported to MRM's OGOR were still much lower than the volumes they reported to BOEMRE's GOADS system and estimates from EPA. Operators reported to OGOR that between 0.3 and 0.5 percent of the natural gas produced on offshore leases was vented and flared each year from 2006 to 2008; however, they reported to GOADS that they vented and flared about 1.4 percent—about 32 Bcf—of the natural gas produced on federal leases in the Gulf of Mexico in 2008.[37] Although regulations require offshore operators to report all sources of lost gas to OGOR, BOEMRE officials said that that this did not include fugitive emissions. Furthermore, these officials also said that operators likely reported volumes from some operational sources as "lease-use" gas instead of including it in the venting and flaring data, thus contributing to the differences between OGOR and GOADS.[38] GOADS data included sources similar to those included in EPA's and WRAP's data for onshore production, including the same operational sources. Further, guidance to operators for reporting to GOADS explicitly outlines the sources to be reported and how they should be estimated, while guidance for OGOR does not. Table 3 outlines the emission sources for volumes operators reported to the GOADS system for 2008. In addition, EPA's offshore estimates showed that around 2.3 percent of gas produced on offshore federal leases—as much as 50 Bcf—was vented and flared every year from 2006 to 2008. According to our analysis of EPA's work, additional venting from natural gas compressors, used to maintain proper pressure in production equipment, accounted for the majority of the difference between the offshore EPA and GOADS volumes.[39]

Table 3. GOADS's Reported Vented and Flared Natural Gas and Sources for Offshore Federal Leases

Sources from 2008 GOADS Study	Volume (Bcf)
Venting	12
Flaring	7
Fugitive emissions	6
Pneumatic devices	3
Glycol dehydrators	1
Other sources	2
Total	32

Source: GAO analysis of GOADS data.
Note: Volume figures in table do not sum to 32 Bcf due to rounding. See appendix I for more information on these figures.

On several occasions BOEMRE has made comparisons between data on vented and flared volumes in the OGOR and GOADS systems, according to BOEMRE officials. In 2004, BOEMRE compared data from the 2000 GOADS study with data from OGOR for a subset of offshore leases and found reported vented and flared volumes were not always in agreement— attributing this difference to different operator interpretations of GOADS and

OGOR reporting requirements. BOEMRE officials said they revised reporting procedures for the 2005 GOADS study. More recently, BOEMRE made similar comparisons between data from the 2008 GOADS study and OGOR data for a subset of leases and found they were in closer agreement.[40] BOEMRE officials told us they will continue to make such comparisons to try to ensure the accuracy of the data in each system. In reporting volumes of vented and flared gas to both systems, operators can choose from a broad array of software packages, models, and equations to estimate emissions, and these techniques can yield widely varied results. For example, one study found that various estimation techniques to determine emissions from oil storage tanks either consistently underestimated or overestimated vented volumes.[41] OGOR reporting instructions for both onshore and offshore operators, as noted, do not specify how operators should estimate these volumes.

As part of our review, we analyzed 2008 OGOR and GOADS data for the Gulf of Mexico and found that the OGOR data likely underestimated the volumes of vented and flared natural gas on federal offshore leases. To do this analysis, we compared 2008 data from GOADS's vent and flare source categories with OGOR data for the same categories—looking at these source categories allowed us to directly compare the two data systems. In doing this analysis, we accounted for OGOR's exclusion of fugitive emissions and the reporting of sources, like pneumatic valves, as lease-use gas. Our analysis found that the volumes operators reported to OGOR— about 12 Bcf—were much lower than the volumes operators reported to GOADS—about 18 Bcf. Neither we nor MRM and BOEMRE officials could account for or explain these differences in the two data systems. BOEMRE officials said that they are still working to improve reporting to OGOR and GOADS and expect these two data systems to converge in the future.

To improve reported data, BOEMRE recently released a final rule, in response to the recommendations in our 2004 report, that requires operators on larger offshore platforms to route vented and flared gas from a variety of sources through a meter to allow for more accurate measurement, among other things.[42] BOEMRE officials said that these meters would help to improve the accuracy of data reported to both OGOR and GOADS.[43] However, BOEMRE officials said they have had to address questions from some operators who were not sure which sources of vented gas should be routed through the newly required meters. In this regard, these officials said it may be useful to enumerate the required emission sources for reporting to OGOR in future guidance to offshore operators. They also noted that BOEMRE is planning a workshop in October 2011 to stress to operators the need for accurate reporting on their submissions to both GOADS and OGOR systems. In a similar way, EPA has taken action to improve the reporting of emissions from the oil and natural gas industry. EPA recently proposed a greenhouse gas reporting rule that would require oil and gas producers emitting over 25,000 metric tons of carbon dioxide equivalent to submit detailed data on vented and flared gas volumes to allow EPA to better understand the contribution of venting and flaring to national greenhouse gas emissions. For onshore leases, the proposed EPA rule provides details on the specific sources of vented and flared gas to be measured and proposes standardized methods for estimating volumes of greenhouse gas emissions where direct measurements are not possible. For offshore leases, operators would use the GOADS system to report venting and flaring. Data collection would begin in 2011 if the rule becomes finalized in 2010.

AVAILABLE TECHNOLOGIES COULD REDUCE ABOUT 40 PERCENT OF NATURAL GAS ESTIMATED TO BE LOST TO VENTING AND FLARING ON ONSHORE FEDERAL LEASES, POTENTIALLY INCREASING FEDERAL ROYALTY PAYMENTS AND REDUCING GREENHOUSE GAS EMISSIONS

Data from EPA, supported by information obtained from technology vendors and our analysis of WRAP data, suggest that about 40 percent of natural gas estimated to be vented and flared on federal onshore leases could be economically captured with currently available control technologies, although some barriers to their increased use exist. Such captures could increase federal royalty payments and reduce greenhouse gas emissions.

Available Technologies Could Reduce Venting and Flaring on Onshore Federal Leases, but According to EPA Officials and Technology Vendors, Some Barriers Exist

Available technologies could reduce venting and flaring at many stages of the production process. However, there are some barriers to implementing these technologies.

Available Technologies Could Reduce Venting and Flaring on Onshore Federal Leases

EPA analysis and our analysis of WRAP data identified opportunities for expanded use of technologies to reduce venting and flaring. Specifically:

- EPA's 2008 analysis, the most recent data available, indicates that the increased use of available technologies, including technologies that capture emissions from sources such as well completions, liquid unloading, or venting from pneumatic devices,[44] could have captured about 40 percent—around 50 Bcf—of the natural gas EPA estimated was lost from onshore federal leases nationwide.[45] For instance, EPA found significant opportunities to add "smart" automation to existing plunger lifts, which tune plunger lifts to maximum efficiency and, in turn, minimize the amount of gas lost to venting. EPA estimated that using this technology where economically feasible could have resulted in the capture of more than 7 Bcf of vented and flared natural gas on federal leases in 2008—around 6 percent of the total volume estimated by EPA to be vented and flared on onshore federal leases. Similarly, EPA estimated that additional wells on onshore federal leases could have incorporated reduced emission completion technologies in 2008, which could have captured an additional 14.7 Bcf of vented and flared natural gas. Table 4 outlines EPA's estimates of potential reductions in venting and flaring on onshore federal leases.

Table 4. Potential Nationwide Reductions on Onshore Federal Leases from Increased Use of Venting and Flaring Reduction Technologies, 2008

Emission source	Potential reduction (Bcf)	Percent of totalvolume EPA estimatedvented and flared
Gas well liquid unloading	7.2—expand use of smart automated plungers	5.7
Well completions	14.7—expand use of reduced emission completions	11.7
Glycol dehydrators	5.7—install vapor recovery devices	4.5
Pneumatic devices	9.7—use low-bleed devices	7.7
Oil and condensate storage tanks	12.9—install vapor recovery units	10.2
Total	50.2	39.8

Source: GAO analysis of EPA data.

- Our analysis of WRAP's 2006 data from certain onshore production basins also highlighted the possibility for additional venting and flaring reductions. We found significant regional differences in use of control technologies in oil and gas production basins in the mountain west and, subsequently, differences in vented and flared volumes, as a percentage of total basin production. For example, according to the WRAP data, most pneumatic devices in the Piceance basin in northwest Colorado were low-bleed in 2006, while high-bleed pneumatic devices were still predominant in the neighboring Uinta basin in northeast Utah. Although the Piceance and Uinta basins are part of the same geological formation and share many characteristics, including type of gas development and extraction methods, the WRAP data show that the venting and flaring volumes on federal leases in the Uinta basin are nearly double those in the Piceance basin as a percentage of total gas production. See appendix I for a map of these basins.

Differing rates of use of venting and flaring reduction technologies— among states, oil and gas production basins, and individual operators— can be attributed to two key factors based on our analysis. Variations in state air quality regulations are one factor, according to EPA and state agency officials and industry representatives. For example, Colorado has stricter requirements than Utah for emissions controls, according to officials, which partly explains the variation in levels of control technologies in use in these production basins. Similarly, over the last 5 years Wyoming has instituted many regulatory changes to address increases in ground-level ozone, including stricter emission-reduction requirements for storage tanks and reduced emission completions.

The net economic benefit of installing equipment to capture vented and flared gas is the second key factor. The cost of implementing a given technology can be substantial—and may be especially burdensome for smaller operators. Nonetheless, in many cases, the costs are recovered quickly as the captured gas is sold, according to industry and EPA officials. According to documents from EPA's Natural Gas STAR program, a voluntary partnership with industry to encourage reductions in gas venting, in many cases, the cost of implementing these control technologies can be recovered in less than 1 year. According to EPA and industry representatives, for operators with sufficient resources— including engineering and cost-estimation teams, as well as capital for infrastructure—decisions to potentially install capture equipment are easy to make based on simple economic considerations. For example, the cost of switching from high-bleed to low-bleed pneumatic devices ranges from $700 to $3,000 per device, which can be recovered in 2 to 8 months, on average, according to EPA documents. Similarly, retrofitting an oil production storage tank with a vapor recovery unit can cost tens of thousands of dollars, but the gas saved can pay for the technology generally within 2 years. According to EPA and BLM officials, some operators have already implemented plunger-lift systems, vapor recovery units, reduced emission completions, and other technologies due to the economic benefit of doing so. However, these officials cautioned that the return-on-investment calculation can be complicated by a number of factors, including the geology and location of the production basin and the differences in the composition of extracted oil or gas. For instance, while some high-bleed pneumatic devices vent more gas than low-bleed devices, the higher bleed rates keep the equipment from freezing in cold conditions, according to industry officials. Similarly, reduced emission completions are not economically viable for conventional gas wells with low wellhead pressures, as the costs of reduced emission completion equipment can easily outweigh the benefits of capturing the gas, according to industry representatives. In addition, EPA and industry officials told us that installing these technologies may require other significant infrastructure investments, such as a new pipeline from an oil well where natural gas is currently being vented or flared to a gas sales line, which could make the investment in these technologies cost-prohibitive.

A number of industry representatives and EPA officials noted the potential for currently-developing carbon markets to influence the economics of venting and flaring control technologies. Carbon markets generally refer to real financial markets where carbon emission reductions, known as carbon offset credits, are bought and sold; companies that are emitting more than the amount of carbon allocated, or allowed by the government regulators, can buy credits to offset their excess emissions and companies that reduce emissions can sell those reductions as credits. Operators are increasingly able to document the carbon reductions achieved through installations of the technologies and, in turn, sell these offset credits on open carbon markets according to industry officials. Potential opportunities to claim and sell these carbon offset credits may add to the economic incentives for using these control technologies, according to some industry officials.[46] Although there is some risk involved with claiming these offsets, as the markets are developing, pending federal legislation could make it more likely that a market for carbon will become increasingly relevant, according to industry officials.

The experiences of some operators in implementing technologies to reduce venting and flaring show the economic advantages that can result. For example, BP installed smart-automated plunger lifts on its onshore wells throughout the San Juan Basin and reported achieving a 99 percent reduction in vented volumes as a result, increasing production and profits, according to company representatives. In addition, according to company officials, the company has replaced all of its high-bleed pneumatic devices with low-bleed devices, saving approximately 3.4 Bcf of natural gas emissions annually. BP also reported investing $1.2 million in reduced emission completions since 2000, which it credits with saving 1.5 Bcf of gas and almost 29,000 barrels of condensate. These emissions savings prevented over 100,000 metric tons of CO_2 and 2,000 metric tons of methane from entering the atmosphere and increased revenues by almost $5.8 million, according to company documents. The results of BP's actions were likely a factor in the estimates of venting and flaring based on WRAP data for the North San Juan basin that we reported earlier; BP was the major operator in that basin, and it had the lowest estimates of venting and flaring. Similarly, Devon Energy recently took steps to expand its use of venting and flaring reduction technologies for some of its onshore wells that have resulted in significant successes, according to company representatives. In 2008, Devon representatives reported 10.4 Bcf of methane emission reductions that they attribute to the replacement of high-bleed pneumatic devices, installation of vapor recovery units on storage tanks, use of automated plunger lift systems, and use of reduced emission completions, among other technologies. Overall, the company saved more than $125 million since 1990 by implementing these technologies, according to Devon representatives.

Vendors of these technologies also cited success stories. For example, one vendor cited an example where installation of two vapor recovery units onshore requiring capital investments of more than $20,000 yielded a full return on investment in less than a month through the capture of otherwise vented gas. In addition, one operator replaced and retrofitted 400 high-bleed pneumatics on wells onshore, at a cost of more than $118,000, but found an annual savings in captured gas of nearly $149,000, for a payback on investment in less than 1 year. Each of these cases demonstrates that the venting and flaring reduction technologies can, under the right circumstances, add to an operator's bottom line. These results illustrate the potential multiple benefits of venting and flaring reduction technologies—benefits to industry in the form of additional revenues; benefits to the government in the form of increased royalty payments; and benefits to the environment in the form of reductions in vented and flared greenhouse gases.

EPA Officials and Technology Vendors Identified Barriers to Implementing Available Technologies

Despite the potential economic benefits of using these technologies, there are barriers to their implementation for some operators, according to an EPA official and technology vendors. One key barrier is that many operators are unaware of the economic advantages. In part, this is because smaller operators often do not have the time or expertise to undertake the engineering analysis to understand whether and how they can benefit, according to EPA and technology vendors. Also, these officials said that smaller operators often do not have the capital to purchase equipment, regardless of whether they can recover the costs. According to officials, the voluntary nature of the EPA Natural Gas STAR program is not enough to spur industry to change, and one industry official stated that the sometimes contentious relationship between the federal government and private industry contributes to this lack of awareness. Private industry does not always take federal efforts to encourage industry to alter business practices at face value, according to officials. One industry representative cited reluctance to participate in EPA's Natural Gas STAR program as an example of this skepticism.

A number of other factors can also contribute to operators not adopting venting and flaring reduction technologies. Officials that we spoke with said that overcoming "institutional inertia"—a company's tendency to do business and carry out operations as it always has—is key to adopting these technologies. In a similar vein, industry and EPA officials told us that upper management support is critical for these types of efforts to go forward, and many companies' management is focused on other efforts that are deemed more important than what are seen as incremental improvements in operations. For example, the operator may choose to invest its limited available capital in drilling a new well, which may have a larger return than investments in capturing vented or flared gas from an existing well, according to industry representatives.

Reductions in Natural Gas Lost to Venting and Flaring Could Increase Federal Royalty Payments and Reduce Greenhouse Gas Emissions

Reductions in natural gas lost to venting and flaring from federal leases would increase the volume of natural gas produced and sold, thereby potentially increasing federal royalty payments. If, for instance, a total of 126 Bcf of natural gas was lost to venting and flaring on onshore federal leases in 2008, as EPA has estimated, that loss would equal approximately $58 million in federal royalty payments. If, as EPA estimates, 40 percent of this lost gas could have been economically captured and sold, federal royalty payments could increase by approximately $23 million annually, which represents about 1.8 percent of annual federal royalty payments on natural gas.[47]

Reducing natural gas lost to venting and flaring from federal leases could also reduce greenhouse gases to the atmosphere according to our calculations. Because methane is about 25 times more potent as a greenhouse gas over a 100-year period, and almost 72 times more potent over a 20-year period according to the Intergovernmental Panel on Climate Change,[48] reducing direct venting of natural gas to the atmosphere has a significantly greater positive effect, in terms of global warming potential, than does reducing flaring. Again using EPA's estimates, if a total of 98 Bcf of natural gas was vented and 28 Bcf was flared annually, those releases would account for about 41 million metric tons of carbon dioxide equivalent released to the atmosphere, which would be roughly equivalent to the emissions of almost 8 million passenger vehicles or about 10 average-sized coal-fired power plants. Capturing 40 percent of this volume would result in emissions reductions of about 50 Bcf, which is equivalent to the emissions of 3.1 million passenger vehicles or about 4 average-sized coal-fired power plants, according to our analysis.[49]

Some EPA officials also told us that they believed that federal efforts to reduce venting and flaring could also have a spillover effect—that is, it could lead operators to use these technologies on state and private leases as well. Data from EPA and WRAP included vented and flared gas from nonfederal leases, and the data showed that there were similar percentages of gas being lost, suggesting that the potential greenhouse gas reductions from the expanded use of these technologies could go well beyond those from federal oil and gas production.[50]

We did not find complete quantitative data on reduction opportunities offshore from Interior, EPA, or others that could be used to fully identify the potential to reduce emissions offshore. However, EPA officials told us that opportunities for reducing emissions from venting and flaring from offshore production platforms likely exist. For instance, EPA found that various production components, including valves and compressor seals, contribute significant volumes of fugitive emissions, but that these emissions could be mitigated through equipment repair or retrofitting. One estimate based on EPA analysis of 15 offshore platforms in 2008, suggests that most of the gas lost through compressor seals could be recovered economically—saving about 70 percent of the overall gas they estimated to be lost on those platforms. However, EPA's analysis warns that some mitigation strategies may be less cost-effective in the offshore environment because capital costs and installation costs tend to be higher.

INTERIOR'S OVERSIGHT DOES NOT ENSURE THAT OPERATORS MINIMIZE VENTING AND FLARING ON FEDERAL LEASES, WHILE A VOLUNTARY EPA PROGRAM HAS REDUCED VENTED GAS ACCORDING TO EPA AND INDUSTRY PARTICIPANTS

Interior is responsible for ensuring that operators minimize natural gas venting and flaring on federal onshore and offshore leases; however, while both BLM and BOEMRE have taken steps to minimize venting and flaring on federal leases, their oversight of such leases has several limitations. Although EPA does not have a direct regulatory role with respect to managing federal oil and gas leases, its Natural Gas STAR program has helped to reduce vented gas on federal leases according to EPA and industry participants.

BLM and BOEMRE Have Taken Steps to Minimize Venting and Flaring on Federal Leases, but Their Oversight Has Several Limitations

As part of their oversight responsibilities, Interior's BLM and BOEMRE are charged with minimizing the waste of federal resources, and, to that end, both agencies have issued regulations and guidance that limit venting and flaring of gas during routine procedures such as liquid unloading and well completions.[51] However, their oversight has several limitations, namely (1) the regulations and guidance do not address new capture technologies or all sources of lost gas; (2) the agencies do not assess options for reducing venting and flaring in advance of oil and gas production for purposes other than addressing air quality; and (3) the agencies have not developed or do not use information regarding available technologies that could reduce venting and flaring.

Regulations and Guidance Limit Venting and Flaring, but Do Not Address Newer Technologies or All Sources of Lost Gas

Onshore Leases

BLM's guidance limits venting and flaring from routine procedures and requires operators to request permission to vent and flare gas above these limits.[52] If operators request permission to exceed these limits, BLM is to assess the economic and technical viability of capturing additional gas and require its capture when warranted.[53] Although BLM guidance sets limits on venting and flaring of natural gas and allows flexibility to exceed them in certain cases, it does not address newer technologies or all sources of lost gas. Specifically, BLM guidance is 30 years old and therefore does not address venting and flaring reduction technologies that have advanced since it was issued. For example, since the guidance was written, technologies have been developed to economically reduce emissions from well completions and liquid unloading—namely the use of reduced emission completion and automated plunger lift technologies respectively. These two sources of emissions were important contributors to vented and flared volumes that we discussed earlier. Despite this fact, the use of such technologies where it is economic to do so is not covered in BLM's current guidance. In general, BLM officials said that they thought the industry would use venting and flaring reduction technologies if they made economic sense. Similarly, new lower-emission devices could also reduce venting and flaring from other sources of emissions that are not covered by BLM's guidance, such as pneumatic valves or gas dehydrators—two sources that contribute to significant lost gas. In discussions with BLM staff about their guidance, staff acknowledged that existing guidance was outdated given current technologies and said that they were planning to update it by the second quarter of 2012.

Offshore Leases

Like BLM, BOEMRE has regulations that limit the allowable volumes of vented and flared gas from offshore leases to minimize losses of gas from routine operations. Operators can also apply for permission to exceed these limits and, like BLM, BOEMRE would evaluate the economic and technical viability of capturing additional gas. Further, BOEMRE inspects offshore platform facilities each year and, as part of these inspections, reviews on-site daily natural gas venting records.[54] BOEMRE officials told us that the agency requires operators to keep these venting records and that it uses them to, among other things, identify any economically viable opportunities for an operator to install control equipment. Overall BOEMRE officials said that operators were required to install venting and flaring reduction equipment where economic, even if they would make as little as $1 in net profit from the captured gas. According to agency officials, due to the type of production and operations offshore, reduction opportunities mostly consist of installing vapor recovery units, and these officials said that they generally believe that companies have installed such equipment where it is economic to do so. Although BOEMRE conducts regular inspections, the daily venting records do not include all sources of vented gas. For example, emission estimates from sources of gas such as pneumatic valves and glycol dehydrators are not included, and therefore inspectors are not able to make assessments of the potential to reduce emissions from these sources. Both of these sources were contributors to lost gas offshore from the 2008 GOADS study, suggesting potential reduction opportunities. BOEMRE officials said that the agency considers these sources lease-use gas, and as a result, believed that they could not legally consider the economic and technical viability of this gas and require its capture when warranted. However, based on our review of BOEMRE regulations and authorizing legislation, it appears that BOEMRE has the authority to require operators to minimize the loss of this gas, including requiring its capture where appropriate. BOEMRE officials agreed with our assessment.

The Agencies Do Not Assess Options for Reducing Venting and Flaring in Advance of Oil and Gas Production

Onshore Leases

While BLM regulations authorize and direct BLM officials to offer technical advice and issue orders for specific lease operations to minimize waste,[55] BLM does not explicitly assess options to minimize waste from vented and flared gas before production. For example, we identified two phases in advance of production where BLM could assess venting and flaring reduction options—during the environmental review phase and when the operator applies to drill a new well. However, the agency does not explicitly assess these options, or discuss them with operators, during either phase. For example, during the environmental review phase, BLM works with states to assess emissions from oil and gas production, and that air quality assessment may include venting and flaring reduction requirements. According to BLM officials, since states generally have primary responsibility to implement and enforce air quality standards, the standards drive these requirements, and states focus only on the role venting and flaring plays in air pollution, rather than the minimization of waste. Therefore in production basins where air quality standards are being met, or where only minimal use of technology is required to meet them, BLM would not assess venting and flaring reduction technologies to the full extent that they could economically reduce vented and flared gas.[56] One official noted that some BLM officials felt constrained in their ability to consider the use of venting and flaring reduction technologies because of this. Similarly, during the phase when operators apply to drill new wells, BLM assesses detailed technical and environmental aspects of the project, but BLM officials told us their assessment does not include a review of options to reduce venting and flaring.[57]

Offshore Leases

Similar to BLM, BOEMRE assesses venting and flaring reduction options in advance of production to determine whether vented and flared gas from offshore platforms would harm coastal air quality, but again, the focus is on meeting air quality standards rather than assessing whether gas can be economically captured. Therefore, when BOEMRE does not anticipate harm to coastal air quality, as is often the case according to officials, the agency does not further consider venting and flaring reduction options at this phase. Further, while the application operators submit in advance of drilling must include a description of the technologies and recovery practices that the operator will use during production,[58] venting and flaring reduction options are not included in that submission.

Agencies Have Not Developed or Do Not Use Information Regarding Available Technologies that Could Reduce Venting and Flaring

Onshore Leases

We found that BLM does not maintain a database regarding the extent to which available venting and flaring reduction technologies are used on federal oil and gas leases.[59] As such, it could be difficult for BLM to identify opportunities to reduce venting and flaring or estimate the potential to increase the capture of gas that is currently vented or flared. For example, while BLM guidance provides that the natural gas vaporizing from storage tanks must be captured if BLM determines recovery is warranted, BLM does not collect data on the use of control technologies and available OGOR data do not contain the volumes of lost gas from storage tanks. Thus BLM may be overlooking circumstances where recovery could be warranted. In addition, according to BLM officials we spoke with, although infrared cameras can be used to identify sources of lost gas, BLM has not used them during inspections of production facilities. Although relatively expensive, infrared cameras allow users to rapidly scan and detect vented gas or leaks across wide production areas. BLM officials cited budgetary constraints and challenges in developing a policy and protocols for why the cameras have not been used regularly by the agency.

Offshore Leases

Although the GOADS data system contains some information on the types of equipment operators use, BOEMRE has not analyzed this information to identify emission-reduction opportunities according to officials. GOADS contains information about the use of equipment such as vapor recovery systems. These data have not been used by BOEMRE to identify venting and flaring reduction opportunities because the agency has not considered using these data for purposes other than addressing air quality, according to a BOEMRE official. Nonetheless, based on our review of the GOADS data system, by not analyzing such data, BOEMRE is not able to identify emission-reduction opportunities. As a case in point, we found that emissions from pneumatic valves in the 2008 GOADS study made noticeable contributions to overall lost gas, which might suggest the potential to expand the use of low-bleed pneumatics in some cases. BOEMRE officials also noted that, unlike BLM, its inspectors had used infrared cameras to look for obvious sources of vented and flared gas in a few sample locations close to shore. In this regard, they said expanded use of infrared cameras could be useful to help enforce their new rule that requires the use of meters for vented and flared gas. Specifically, they said that the cameras could identify sources of gas that operators may have not routed through the meter as required. They also noted that expanded use of the cameras could help to identify and potentially reduce fugitive gas emissions that currently go undetected.[60]

EPA's Voluntary Natural Gas STAR Program Has Helped Reduce Vented Gas, According to EPA and Industry Participants

Although Interior has the primary role in federal oil and gas leasing, EPA's Natural Gas STAR program has encouraged some operators to adopt technologies and practices that have helped to reduce methane emissions from the venting of natural gas, according to EPA and industry participants. Through this program, industry partners evaluate their emissions and consider ways to reduce them, although the reductions are voluntary. The program also maintains an online library of technologies and practices to reduce emissions that quantify the costs and benefits of each emission-reduction option. Natural Gas STAR also sponsors conferences to facilitate information exchange between operators regarding emissions reductions technologies. Partner companies report annually about their efforts to reduce emissions along with the volumes of the emission reductions.[61]

According to the Natural Gas STAR Web site, domestic oil and gas industry partners reported more than 114 Bcf of methane emission reductions in 2008, which amounts to about 0.4 percent of the total natural gas produced that year. However, one industry representative said that, while large and midsize operators were aware of the Natural Gas STAR program, smaller operators were not aware and, even if some smaller operators were aware of the program, they may not have the environmental staff to implement the technologies and practices. Despite the potential usefulness of information from the Natural Gas STAR program to oil and gas producers on federal leases, some of the BLM officials that we spoke with were unfamiliar with Natural Gas STAR.

CONCLUSIONS

Fulfilling its responsibility to ensure that the country's oil and natural gas assets are developed reasonably and result in fair compensation for the American people requires Interior to have accurate and complete information on all aspects of oil and natural gas leases. Interior has collected some information on vented and flared gas through MRM's OGOR system, but without a full understanding of these losses Interior cannot fully account for the disposition of taxpayer resources or identify opportunities to prevent undue waste. MRM's OGOR data system does not provide information on all sources of lost gas, which is the primary source of data that BLM uses to measure overall vented and flared gas onshore. Therefore, OGOR data present an incomplete picture of venting and flaring onshore, leading BLM officials to believe that vented and flared gas volumes do not represent a significant loss of gas on federal leases. Similarly, data in BOEMRE's GOADS data system differ considerably from data in OGOR, and have not been reconciled—raising questions about the accuracy of offshore data sources.

Regarding Interior's oversight of operators venting and flaring gas, because current guidance and regulations from BLM and BOEMRE do not require the minimization of all sources of vented and flared gas—although legislation exists authorizing them to require that waste on federal leases be minimized—operators may be venting and flaring more gas than should otherwise be allowed. In fact, we found that operators are not using available technologies in all cases to economically reduce vented and flared gas. BLM guidance has not kept pace with the development of economically viable capture technologies for a number of sources of lost gas, and BOEMRE has been reluctant to consider the economic and technical viability of minimizing the waste of "lease-use" gas because officials had believed they were legally constrained from doing so.

In addition to the limitations of these regulations, BLM and BOEMRE have not used their authority in two situations where they could potentially further reduce venting and flaring. First, neither agency has used its authority to minimize waste beyond relevant air quality standards by assessing the use of venting and flaring reduction technologies before production. Second, because BLM lacks data about the use of venting and flaring technologies for onshore leases and BOEMRE does not analyze its existing information for offshore leases in its GOADS data system, these agencies are not fully aware of potential opportunities to use available technologies. Further, neither agency takes full advantage of newer infrared camera technology that can help to identify sources of lost gas— as BOEMRE officials have acknowledged, this technology could help reveal additional sources of lost gas.

Ultimately, a sharper focus by BOEMRE and BLM on the nature and extent of venting and flaring on federal leases could have multiple benefits. Specifically, increased implementation of available venting and flaring reduction technologies, to the extent possible, could increase sales volumes and revenues for operators, increase royalty payments to the federal government, and decrease emissions of greenhouse gases. In addition, our analysis of WRAP and EPA data showed as much or more vented and flared gas on nonfederal leases, and we share the observation with EPA officials that a spillover effect may occur, whereby oil and gas producers, seeing successes on their federal leases, take similar steps on state and private leases.

RECOMMENDATIONS FOR EXECUTIVE ACTION

To ensure that Interior has a complete picture of venting and flaring on federal leases and takes steps to reduce this lost gas where economic to do so, we are making five recommendations to the Secretary of the Interior.

To ensure that Interior's data are complete and accurate, we recommend that the Secretary of the Interior direct BLM and BOEMRE to take the following action:

- • Take additional steps to ensure that each agency has a complete and accurate picture of vented and flared gas, for both onshore and offshore leases, by (1) BLM developing more complete data on lost gas by taking into consideration additional large onshore sources and ways to estimate them not currently addressed in regulations—sources that EPA's newly proposed greenhouse gas reporting rule addresses—and (2) BOEMRE reconciling differences in reported offshore venting and flaring volumes in OGOR and GOADS data systems and making adjustments to ensure the accuracy of these systems. To help reduce venting and flaring of gas by addressing limitations in their regulations, we recommend that the Secretary of the Interior direct BLM and BOEMRE to take the following four actions:
- BLM should revise its guidance to operators to make it clear that technologies should be used where they can economically capture sources of vented and flared gas, including gas from liquid unloading, well completions, pneumatic valves and glycol dehydrators. BOEMRE should consider extending its requirement that gas be captured where economical to "lease-use" sources of gas;
- BLM and BOEMRE should assess the potential use of venting and flaring reduction technologies to minimize the waste of natural gas in advance of production where applicable, and not solely for purposes of air quality;
- BLM and BOEMRE should consider the expanded use of infrared cameras, where economical, to improve reporting of emission sources and to identify opportunities to minimize lost gas; and
- BLM should collect information on the extent that larger operators use venting and flaring reduction technology and periodically review this information to identify potential opportunities for oil and gas operators to reduce their emissions, and BOEMRE should use existing information in its GOADS data system for this same purpose, to the extent possible.

AGENCY COMMENTS AND OUR EVALUATION

We provided a copy of our draft report to Interior and EPA for review and comment. Interior provided written comments that concurred with four of the five recommendations and partly concurred with the remaining recommendation. Its comments are reproduced in appendix II and key areas are discussed below. EPA did not provide formal comments on the report, but the agency's Office of Air and Radiation provided written comments to GAO staff, which we summarize and discuss below. Interior and EPA also provided other clarifying or technical comments, which we incorporated as appropriate.

Interior's comments reflected the views of BLM and BOEMRE. BLM concurred with all five recommendations and noted that it plans to incorporate recommended actions into its new Onshore Order in order to improve the completeness and accuracy of its data and help address limitations in its current regulations.

BOEMRE concurred with four of the recommendations and partly concurred with our second recommendation that they consider enforcing the economical capture of "lease-use" gas. It stated that we misapprehended the scope of the regulations governing "lease-use" sources of gas in that BOEMRE does not have current regulations to require the capture of "lease-use" gas. In response to this comment, we reworded our recommendation to clarify that BOEMRE should consider extending its existing requirements for the economical capture of gas to "lease-use" gas. In a related point, BOEMRE also noted that we were unable to quantify the potential volumes of additional gas that could be captured by holding operators to this same economic standard for "lease-use" gas. While current data have limitations, BOEMRE's GOADS data suggest potential opportunities to capture additional gas from lease-use sources, namely glycol dehydrators and pneumatic devices. As such, we support BOEMRE's efforts to further evaluate this issue and take action through new guidance or regulations, as it believes appropriate.

EPA's Office of Air and Radiation commented on three areas of the report:

- First, EPA emphasized the significant air quality impacts from the volatile organic compounds (VOC) associated with vented gas and provided us with estimates of the potential volumes of these emissions. While we recognize that the impacts of VOC emissions on air quality are important, these impacts were largely beyond the scope of our work. Nonetheless, we incorporated an estimate of these VOC emissions into supporting notes to table 1 that reflected EPA's estimates of vented and flared gas. We also added additional information to the background regarding VOC emissions.
- Second, EPA suggested that we recommend to BLM and BOEMRE that they require the use of the best available venting and flaring control measures during leasing or drilling permitting. We continue to believe that BLM and BOEMRE should require the use of these technologies where economical, and recognize that requiring the use of such controls when the economics of capturing gas are unfavorable is not required by current EPA greenhouse gas regulations.
- Third, EPA provided us with its revised emission estimates for vented and flared gas based on updated analysis for its proposed rule on the reporting of greenhouse gases by industry. It also provided us with revised estimates for the use of additional control technologies to reduce the emissions of vented and flared gas. In both cases, we incorporated these revised estimates in our report where applicable.

As agreed with your offices, unless you publicly announce the contents of this report earlier, we plan no further distribution until 30 days from the report date. At that time, we will send copies of this report to the appropriate congressional committees, Secretary of the Interior, Administrator of the Environmental Protection Agency, and other interested parties. In addition, the report will be available at no charge on the GAO Web site at http://www.gao.gov.

If you or your staffs have any questions about this report, please contact me at (202) 512-3841 or ruscof@gao.gov. Contact points for our Offices of Congressional Relations and

Public Affairs may be found on the last page of this report. GAO staff who made major contributions to this report are listed in appendix IV.

Frank Rusco
Director, Natural Resources and Environment

APPENDIX I: OBJECTIVES, SCOPE, AND METHODOLOGY

Our objectives were to (1) examine available estimates of vented and flared natural gas on federal leases; (2) estimate the potential to capture additional vented and flared natural gas with available technologies and the associated potential increases in royalty payments and reductions in greenhouse gas emissions and; (3) assess the federal role in reducing venting and flaring of natural gas.

To examine available estimates of vented and flared natural gas on federal leases, we collected data from the Department of the Interior's (Interior) Bureau of Land Management (BLM), Bureau of Ocean Energy Management, Regulation and Enforcement (BOEMRE), including BOEMRE's Minerals Revenue Management (MRM) program; the Environmental Protection Agency (EPA); and the Western Regional Air Partnership (WRAP). We also interviewed staff from these agencies and oil and gas producers operating on federal leases regarding venting and flaring data collection, analysis, and reporting. We obtained data from four key sources: MRM's Oil and Gas Operations Report (OGOR) database, BOEMRE's Gulfwide Offshore Activity Data System (GOADS), EPA's Natural Gas STAR Program, and WRAP's analysis of air emissions for a number of western states. We assessed the quality of the data from each of these sources and determined that these data were sufficiently reliable for the purposes of our report.

MRM provided OGOR data on vented and flared volumes and production for both onshore and offshore federal leases for calendar years 2006 to 2008. MRM uses the OGOR data, in part, to ensure accurate federal royalty payments.[1] The OGOR data are operator-reported, and reported venting and flaring volumes are a mix of empirical measurements and estimates from operators. MRM was unable to provide complete estimates of vented and flared gas on all federal leases because a portion of federal leases are managed as part of lease agreements—collections of leases that draw from the same oil or gas reservoir, which may include federal and nonfederal leases. MRM was unable to determine the share of reported vented and flared gas from the federal portion of those lease agreements; it reported venting and flaring from (1) lease agreements that included only federal leases and (2) all lease agreements, which included some nonfederal leases. In this report, we discuss the vented and flared volumes from the agreements that contain only federal leases. As a result, we report vented and flared gas volumes from the OGOR data as a percentage of total production on these leases, rather than as absolute volumes, in order to compare the OGOR estimates to estimates from other data sources.

---

[1] Operators report royalties using a separate data collection form that operators are required to report to BOEMRE monthly.

A second source of venting and flaring data was BOEMRE's 2008 GOADS data, which contained estimates of gas lost to venting and flaring on federal leases in the Gulf of Mexico—which accounted for 98 percent of federal offshore gas production in 2008. BOEMRE collects GOADS data every 3 years and uses these data to estimate the impacts of offshore oil and gas exploration, development, and production on onshore air quality in the Gulf of Mexico region. BOEMRE also uses GOADS as part of an impact analysis required by the National Environmental Policy Act. GOADS data capture specific information on a variety of sources of air pollutants and greenhouse gases resulting from offshore oil production. BOEMRE provided us with actual volumes of natural gas released from the vented and flared source categories. For the other sources, we used the emissions that were reported in GOADS in tons of methane per year, and we converted these to volumes of methane and then to natural gas, assuming a 78.8 percent methane content for natural gas.[2] In the GOADS study, fugitive emissions are estimated by looking at the number of valves and other components on a given production platform and then assuming an average leak rate. BOEMRE's data contractor performs a series of quality checks on the data after collection.

A third source of data on vented and flared volumes was a nationwide analysis performed by officials from EPA's Natural Gas STAR program, a national, voluntary program that encourages oil and gas companies, through outreach and education, to adopt cost-effective technologies and practices that improve operational efficiencies and reduce methane emissions. EPA's nationwide venting and flaring volumes were based on publicly available empirical data on national oil and gas production for 2006, 2007, and 2008, combined with knowledge of current industry practices, including usage rates and effectiveness of venting and flaring reduction technologies. For example, EPA used data on the number of well completions per year and data on the average venting per completion to estimate a yearly nationwide total from that source, with similar approaches used for estimating total venting and flaring from other key sources.[3] EPA adjusted its estimates to account for the industry's efforts to control some venting and flaring emissions. EPA's analysis was limited in some ways, however. For instance, lacking empirical data on actual nationwide rates of use of certain control technologies, EPA based its analysis on anecdotal information in some cases. In order to be able to compare these data with the OGOR data, we scaled EPA's national estimates to federal leases based on the proportion of natural gas production on federal leases over total U.S. natural gas production using data from MRM and the Department of Energy's Energy Information Administration (EIA).[4] EPA also made estimates of offshore venting and flaring based on BOEMRE's 2005 GOADS data. EPA officials adjusted volumes reported to GOADS based on publicly available information on current industry practices, including usage rates and effectiveness of venting and flaring reduction technologies.

---

[2] We combined the pneumatic pump and pressure/level controller categories into our "pneumatic devices" category. We combined all the other source categories from GOADS into our "other sources" category.

[3] Due to incomplete data on oil storage tank emissions and reductions for 2008, the tank emissions from 2007 serves as an approximation for the emissions in 2008. EPA also included workovers with its estimates of venting and flaring from well completions. Workovers are remedial procedures designed to increase production on existing oil and gas wells.

[4] EPA's initial estimates of venting and flaring were for the methane component of natural gas. These volumes were converted to reflect overall natural gas emissions by assuming, for most sources, an average 78.8 percent methane content for the gas.

A fourth source of venting and flaring data was based on analysis conducted by WRAP, a collaborative arrangement between tribal and state governments and various federal agencies set up to develop the technical and policy tools needed by western states and tribes to comply with the EPA's regional air quality regulations. As part of its efforts to better understand the oil and gas industry's impact on regional air quality, WRAP, through its contractor, the Environ International Corporation, collected data for 2006 on the volumes and sources of key air pollutants such as volatile organics and nitrogen oxides, which are associated with vented and flared gas. WRAP collected these data with backing from the Independent Petroleum Association of Mountain States, an industry group representing oil and gas producers in the western United States. We used Environ to reconfigure the data from the WRAP air quality analysis in order to estimate the overall volumes of vented and flared gas. The WRAP analysis focused on five specific production basins in the mountain west: the Piceance, Denver-Julesburg, and North San Juan Basins in Colorado; the Uinta Basin in eastern Utah; and the South San Juan Basin in northern New Mexico (see fig. 4). The WRAP analysis was based primarily on empirical data from operators in these basins, including drilling and production volume data, as well as data from a survey of operators. This survey asked operators to report actual vented and flared volumes, as well as to provide information on other aspects of their operations, including the emission control technologies they had in place. Similar to the EPA venting and flaring analysis, however, Environ did not have complete data from all operators in each basin and thus estimated some information based on survey data from a subset of operators.[5] In addition, the original WRAP data did not distinguish between federal and nonfederal oil and gas operations, so we provided federal well numbers to Environ so that they could identify the federal lease component of vented and flared gas.

Source: WRAP/Environ.

Figure 4. Locations of Production Basins Included in WRAP Study.

---

[5] BP provided most of the data for the North San Juan basin, and Environ was not able to verify its accuracy to the extent that it did for data reported in the other basins.

To estimate the magnitude of potential increases in royalty payments and reductions in greenhouse gas emissions resulting from capturing additional vented and flared gas with available technologies, we had EPA provide us with estimates of the onshore expansion potential of a number of key technologies and associated venting and flaring volume reductions. For simplicity, EPA developed these estimates by focusing on the expansion potential of a subset of technologies considered to provide the largest emission reductions. These estimates may be conservative, however, because they did not incorporate reductions from a number of other potential venting and flaring opportunities catalogued by the Natural Gas STAR program.[6] These estimates were not based entirely on comprehensive usage data collected from the oil and gas industry, but were based, in part, on publicly available evidence collected through years of experience with the oil and gas industry. In addition, circumstances are constantly changing, and more technological innovations are potentially being used as time goes on, so there is some uncertainty in how much lost gas can be captured. We also compared venting and flaring volumes and the types of emission-reduction technologies used in each of the basins from the WRAP data, allowing us to draw conclusions about the impact of different levels of technology on venting and flaring volumes. We did not identify similar data on reduction opportunities offshore. We also interviewed officials from BLM, BOEMRE, EPA, and state agencies, as well as representatives from private industry, including technology vendors and an environmental consultant regarding the expanded use of available technologies to capture additional vented and flared gas. We conducted background research on venting and flaring reduction technologies, including from publicly available EPA Natural Gas STAR case studies. Finally, we obtained royalty information from MRM to calculate the royalty implications of the onshore venting and flaring reductions, and used conversion factors from EPA to calculate the greenhouse gas impacts of the vented and flared natural gas.[7]

To assess the federal role in reducing vented and flared gas, we conducted interviews with officials from Interior, EPA, the Department of Energy, state agencies, and members of the oil and gas industry. We also reviewed agency guidance and documentation, other studies related to federal management and oversight of the oil and gas industry, as well as prior GAO work that described limitations in the systems Interior has in place to track oil and gas production on federal leases.[8] We conducted interviews with officials in six BLM field offices (Farmington and Carlsbad in New Mexico; Vernal, Utah; Glenwood Springs, Colorado; Pinedale, Wyoming; and Bakersfield, California) and staff from BLM headquarters. We also interviewed BOEMRE staff in Denver, Colorado, and New Orleans, Louisiana.

We conducted this performance audit from July 2009 to October 2010 in accordance with generally accepted government auditing standards. Those standards require that we plan and

---

[6] It is difficult to estimate the extent to which the use of each control technology can be increased. Reductions may not always be feasible and depend on site-specific conditions. EPA's estimates of potential reductions from oil and condensate storage tanks also involved valve inspection and repair in addition to installing vapor recovery units.

[7] The conversion factor we used was .4045 million metric tons of carbon dioxide equivalent per billion cubic feet of vented natural gas, and 0.06 million metric tons of carbon dioxide per billion cubic feet of flared natural gas. We used a royalty rate of 11.45 percent and an average natural gas price of $4.01 per thousand cubic feet.

[8] GAO, Oil and Gas Management: Interior's Oil and Gas Production Verification Efforts Do Not Provide Reasonable Assurance of Accurate Measurement of Production Volumes, GAO-10-313 (Washington D.C.: Mar. 15, 2010).

perform the audit to obtain sufficient, appropriate evidence to provide a reasonable basis for our findings and conclusions based on our audit objectives. We believe that the evidence obtained provides a reasonable basis for our findings and conclusions based on our audit objectives.

APPENDIX II: COMMENTS FROM THE DEPARTMENT OF THE INTERIOR

United States Department of the Interior

OFFICE OF THE SECRETARY
WASHINGTON, D.C. 20240

OCT 12 2010

TAKE PRIDE
IN AMERICA

Mr. Frank Rusco
Director, Natural Resources and Environment
Government Accountability Office
441 G Street, N.W.
Washington, D.C. 20548

Dear Mr. Rusco:

Thank you for the opportunity to review and comment on the Government Accountability Office (GAO) draft report entitled, *Federal Oil and Gas Leases: Opportunities Exist to Capture Vented and Flared Gas, Which Would Increase Royalty Payments and Reduce Greenhouse Gases* (GAO-11-34). The draft GAO report includes five recommendations for the Secretary of the Interior that are intended to augment the present capability to capture vented and flared gas, potentially increase royalty payments, and reduce greenhouse gas emissions. More specifically, the GAO's recommendations address emissions data collection; the reconciliation of conflicting data; providing new emission reporting guidance to operators; and the need to explore using new technologies to identify and reduce volumes of flared and vented natural gas.

The Department of the Interior (DOI) concurs or partially concurs with all five recommendations. Within DOI, the Bureau of Land Management (BLM) and Bureau of Ocean Energy Management, Regulation and Enforcement (BOEMRE) have the major responsibilities for leasing Federal oil and gas and accounting for fluid mineral revenues. Responses to each recommendation are provided in the attached Enclosure. In addition, for your consideration, technical comments are being provided in a separate electronic transmission.

The DOI acknowledges GAO's concerns about the potential impacts from uncaptured vented and flared gas. As noted in your draft report, both BLM and BOEMRE took steps to minimize venting and flaring on Federal oil and gas leases following the 2004 GAO report on BLM and BOEMRE venting and flaring processes (GAO-04-809). Since the 2004 GAO report, both BLM and BOEMRE have issued guidance and regulations to limit venting and flaring of gas during routine operations. Current initiatives demonstrate the Department's continued efforts to improve its monitoring and reduction of vented and flared gas. For example, BLM is revising Onshore Orders, which will address use of new technology to reduce gas emissions. The BOEMRE published a final rule in the *Federal Register* on April 19, 2010, requiring operators on larger offshore platforms to route vented and flared gas from a variety of sources through a meter. This will enable operators to report flared gas separately from vented gas and allow better tracking of greenhouse gas emissions. Additionally, both BOEMRE and BLM will track the Environmental Protection Agency's greenhouse gas rulemaking and use venting and flaring reduction technology where it is economic.

As GAO reported in 2004, the Federal lands and waters already have among the best venting and flaring records in the world. The DOI is committed to managing venting and flaring from onshore and offshore Federal oil and gas leases to the extent of our authority and will continue to improve our programs to remain a world leader in oversight of venting and flaring.

We appreciate your suggestions for improving the regulatory oversight of Federal oil and gas. If you have any questions, please contact Andrea Nygren, BOEMRE Audit Liaison Officer, at 202-208-4343, or LaVanna Stevenson-Harris, BLM Audit Liaison Officer, at 202-912-7077.

Sincerely,

Wilma A. Lewis
Assistant Secretary
Land and Minerals Management

Enclosure

Enclosure

DOI Response to Government Accountability Office (GAO) Draft Report
Federal Oil and Gas Leases: Opportunities Exist to Capture Vented and Flared Gas, Which Would Increase Royalty Payments and Reduce Greenhouse Gases (GAO-11-34)

To help ensure that Interior's data is complete and accurate, the GAO recommends that the Secretary of the Interior direct the Bureau of Land Management (BLM) and the Bureau of Ocean Energy Management, Regulation and Enforcement (BOEMRE) to:

Recommendation 1: *Take additional steps to ensure that each agency has a complete and accurate picture of vented and flared gas, for both onshore and offshore leases, by (1) BLM developing more complete data on lost gas by taking into consideration additional large onshore sources and ways to estimate them not currently addressed in regulations – sources that EPA's newly proposed greenhouse gas reporting rule addresses and (2) BOEMRE reconciling differences in reported offshore venting and flaring volumes in OGOR and GOADS data systems and making adjustments to ensure accuracy of these systems.*

BLM: Concurs
BOEMRE: Concurs

The BLM will track the content and timing of the Environmental Protection Agency's (EPA) greenhouse gas rulemaking. The BLM will analyze options for new standards to track additional sources of natural gas released during drilling, testing, and production. The new standards will be incorporated in the new Onshore Order on waste prevention and beneficial use.

In the current BOEMRE environment, it is difficult to compare the volumes of gas vented and flared in Oil and Gas Operations Reports (OGOR) to the Gulfwide Offshore Activity Data System (GOADS), but reconciling the database differences may be feasible in the future. The functions and structure of the BOEMRE OGOR and the GOADS differ. The OGOR is a lease-based report that historically did not require vented and flared gas volumes to be separately reported. The GOADS is facility-based, and data collected and reported to GOADS separates venting and flaring volumes. In May 2010, the agency published new regulations requiring companies to install flare/vent meters on facilities that process more than 2,000 barrels of oil per day (BOPD), on average. The regulations also require all venting and flaring to be reported separately on the OGOR (even facilities processing less than 2,000 BOPD). As recommended in the 2004 GAO report (GAO-04-809), the Minerals Management Service (now BOEMRE) performed a cost-benefit analysis to determine the proper threshold at which companies should be required to use venting and flaring meters based on economic feasibility. The agency determined that it would be economically feasible for facilities that meet the 2,000 BOPD threshold. We expect that the new regulations requiring meters on these facilities will allow us to compare facility volumes reported on the OGOR and GOADS system, reconcile data on larger facilities, and improve the accuracy of data in both systems.

An important component for improving data accuracy will be educating the operators and reporters collecting and entering data into each system. The BOEMRE recently issued a Notice to Lessees (NTL) and Operators in the Gulf of Mexico announcing the GOADS 2011 effort. The NTL provided a GOADS users' guide and answers Frequently Asked Questions. It also announced a GOADS workshop, scheduled for October 2010, to discuss and explain the information collection and reporting procedures. In addition, the BOEMRE's Minerals Revenue Management program—now the Office of Natural Resources Revenue (ONRR)—issued a Dear Reporter letter in May 2010 to all operators, instructing them on how to report flaring and venting separately on the OGOR. The ONRR also conducts an average of four reporter training sessions a year. The ONRR will continue to educate operators on the flaring and venting reporting requirements, and in particular the new regulations, disposition codes, and meters needed to report correctly. This training will emphasize the need to accurately report flaring and venting on the OGOR and in the GOADS systems. The next reporter training will be held in February 2011. This will be a joint effort between the BOEMRE offshore and revenue management programs.

Once sufficient data are available and analyzed, BOEMRE can determine if additional reconciliation efforts are needed.

To help reduce venting and flaring of gas addressing limitations in its regulations [GAO] recommend[s] that the Secretary of the Interior direct BLM and BOEMRE to take the following four actions:

Recommendation 2: *BLM should revise its guidance to operators to make it clear that technologies should be used where they can economically capture sources of vented and flared gas, including gas from liquid unloading, well completions, pneumatic valves, and glycol dehydrators. BOEMRE should consider enforcing its requirement that gas be captured where economical for "lease-use" sources of gas.*

BLM: Concurs
BOEMRE: Partially concurs

The BLM will develop new standards to require use of new technologies that can economically capture vented and flared natural gas used in lease operations. The new standards will be incorporated in the new proposed Onshore Order on waste prevention and beneficial use.

With respect to BOEMRE, the GAO's recommendation 2 misapprehends the scope of the regulations governing "lease-use" sources of gas. Because BOEMRE has no present requirement to capture "lease-use" sources of gas where economic, new regulations would have to be implemented before such a requirement could be imposed and enforced. As mentioned in your report, GAO was unable to quantify the volume of natural gas that would be saved by such a regulatory change. It is highly uncertain at this time if the volume of gas saved would be large enough to make a capture requirement economic. Thus, BOEMRE will evaluate the issue further and determine if new regulations are warranted.

Recommendation 3: *BLM and BOEMRE should assess the potential use of venting and flaring reduction technologies to minimize the waste of natural gas in advance of production where applicable, and not solely for purposes of air quality.*

BLM: Concurs
BOEMRE: Concurs

The BLM will analyze options for new standards to require use of venting and flaring reduction technologies, where applicable. The new standards will be incorporated in the new Onshore Order on waste prevention and beneficial use.

The BOEMRE diligently addresses air quality in advance of production. The meter requirement stemming from new regulations will provide data on the extent of venting and flaring. Once sufficient data exist, BOEMRE will use the data collected from the new meters to analyze the need and feasibility of requiring venting and flaring reduction technologies on facilities in advance of production, for purposes other than air quality.

Recommendation 4: *BLM and BOEMRE should consider the expanded use of infrared cameras where economic to improve reporting of emission sources and to identify opportunities to minimize lost gas.*

BLM: Concurs
BOEMRE: Concurs

The BLM agrees that infrared cameras can be a useful tool for detecting natural gas emissions as part of a directed inspection program. The BLM will implement the use of infrared cameras during inspections to spot check emission sources and minimize lost gas.

The BOEMRE will expand its use of infrared cameras where economic.

Recommendation 5: *BLM should collect information on the extent that larger operators use venting and flaring reduction technology and periodically review this information to identify potential opportunities for oil and gas operators to reduce their emissions, and BOEMRE should use existing information in its GOADS data system for this same purpose, to the extent possible.*

BLM: Concurs
BOEMRE: Concurs

The BLM will analyze options to collect information on how larger operators use venting and flaring reduction technology on Federal onshore leases, and periodically review this information to identify potential opportunities for oil and gas operators to reduce their emissions. The options adopted will be incorporated in the new Onshore Oil and Gas Order on waste prevention and beneficial use.

The BOEMRE will use the meters required by the new regulations to gather data on offshore venting and flaring volume, and periodically review and identify potential opportunities for oil and gas operators to reduce their emissions. The BOEMRE will also evaluate the 2008 data in GOADS to identify potential opportunities where reduction technology could be economically applied under the Outer Continental Shelf Lands Act conservation mandate. The BOEMRE will track EPA's greenhouse gas rulemaking and use venting and flaring reduction technology where it is economic.

APPENDIX III: VOLUMES AND SOURCES OF VENTED AND FLARED GAS BASED ON ANALYSIS OF 2006 WRAP DATA

Table 5. Volumes and Sources of Vented and Flared Gas from the Uinta Basin

Sources from Uinta Basin	Volume (Bcf)
Pneumatic devices	4.3
Glycol dehydrators	4.3
Fugitive emissions	0.4
Oil and condensate storage tanks	0.2
Other sources	0.1
Total	9.2

Source: Environ Corp. analysis of 2006 WRAP data.
Note: Volume figures in table may not sum to totals due to rounding.

Table 6. Volumes and Sources of Vented and Flared Gas from the North San Juan Basin

Sources from North San Juan Basin	Volume (Bcf)
Glycol dehydrators	0.053
Gas well liquid unloading	0.004
Flaring	0.003
Pneumatic devices	0.001
Fugitive emissions	0.001
Total	0.062

Source: Environ Corp. analysis of WRAP data.
Note: Volume figures in table may not sum to totals due to rounding.

Table 7. Volumes and Sources of Vented and Flared Gas from the South San Juan Basin

Sources from South San Juan Basin	Volume (Bcf)
Well completions	5.3
Glycol dehydrators	3.2
Gas well liquid unloading	1.9
Fugitive emissions	0.6
Pneumatic devices	0.4
Other sources	0.1
Total	11.6

Source: Environ Corp. analysis of WRAP data.
Note: Volume figures in table may not sum to totals due to rounding.

Table 8. Volumes and Sources of Vented and Flared Gas from the Denver-Julesburg Basin

Sources from Denver-Julesburg Basin	Volume (Bcf)
Oil and condensate storage tanks	0.035
Pneumatic devices	0.030
Fugitive emissions	0.019
Gas well liquid unloading	0.006
Well completions	0.002
Other sources	0.004
Total	0.095

Source: Environ Corp. analysis of WRAP data.
Note: Volume figures in table may not sum to totals due to rounding.

End Notes

[1] In June 2010, the Secretary of the Interior changed the name of the Minerals Management Service to BOEMRE.

[2] For the purposes of this report, we use the term "natural gas" to mean the mixture of gas resulting from oil and gas production activities. This natural gas will vary in content but, on average, is approximately 80 percent methane, with the remaining 20 percent a mix of other hydrocarbons and nonhydrocarbons, such as carbon dioxide and nitrogen.

[3] Major greenhouse gases include carbon dioxide (CO_2); methane (CH_4); nitrous oxide (N_2O); and synthetic gases such as hydrofluorocarbons (HFC), perfluorocarbons (PFC), and sulfur hexafluoride (SF_6).

[4] GAO, *Natural Gas Flaring and Venting: Opportunities to Improve Data and Reduce Emissions*, GAO-04-809 (Washington D.C.: July 14, 2004).

[5] Much of our information and data about offshore leases from BOEMRE came from its Offshore Energy and Minerals Management program.

[6] EPA developed these estimates to support a proposed rule in 2010 to require the reporting of greenhouse gas emissions from the oil and gas industry. These estimates were for 2006 to 2008.

[7] WRAP is a collaborative effort of tribal governments, state governments, and various federal agencies to address western air quality concerns. It is administered by the Western Governors' Association and the National Tribal Environmental Council.

[8] Independent Petroleum Association of Mountain States is an industry association representing oil and gas producers in the western United States.

[9] Production basins are land formations with subsurface oil and natural gas reservoirs, often covering hundreds of square miles.

[10] GAO, *Oil and Gas Management: Interior's Oil and Gas Production Verification Efforts Do Not Provide Reasonable Assurance of Accurate Measurement of Production Volumes*, GAO-10-313 (Washington D.C.: Mar. 15, 2010).

[11] Pub. L. No. 66-146, 41 Stat. 437 (1920).

[12] 67 Stat. 462 (1953) codified at 43 U.S.C. § 1331 *et seq.*

[13] These routine amounts are laid out in BLM and BOEMRE regulations and guidance and allow certain volumes from a number of operations, such as well completions. Operators are required to notify these agencies if they plan to intentionally vent and flare beyond routine amounts. BLM and BOEMRE then classify the gas as either unavoidably or avoidably lost based on their judgment of the technical or economic feasibility of capturing the gas, and royalties are due on losses deemed avoidably lost.

[14] While not considered "vented" gas by Interior, we include fugitive emissions in this report as a relevant source of lost gas.

[15] Condensates are hydrocarbon liquids that are byproducts of natural gas production. Unprocessed natural gas normally contains a small amount of water vapor, and dehydration equipment is used to remove this moisture prior to pipeline transportation.

[16] Older detection technology consisted of using manual probes, which need to be very close to the venting source.

[17] This video shows vented gas, which appears to be smoke, billowing from the top of cylindrical metal oil storage tanks and from a pneumatic valve. Video clips courtesy of EPA and a private emission detection firm.

[18] The meters used by BOEMRE and BLM to establish production volumes, upon which royalty payments are based, are generally located downstream of the production site, at the point where the oil and gas enter the sales pipeline or other distribution network.

[19] The Intergovernmental Panel on Climate Change is the body for the assessment of climate change, established by the United Nations Environment Programme and the World Meteorological Organization to provide a scientific view on the current state of climate change and its potential environmental and socio-economic consequences.

[20] EPA defines volatile organic compounds as certain carbon compounds that participate in atmospheric chemical reactions.

[21] EPA is responsible for the regulation of such air pollutants and has been recently charged with regulating greenhouse gas emissions under the Clean Air Act.

[22] BOEMRE has primary regulatory responsibility over air emissions from offshore sources in the central and western Gulf of Mexico.

[23] See EPA's Natural Gas STAR Web site for more information on these technologies (www.epa.gov/gasstar/).

[24] Triethylene glycol is the active chemical in the operation of this equipment. A flash tank separator is a device that captures additional gas from glycol dehydrators.

[25] GAO-04-809.

[26] BLM determined that requiring thousands of onshore operators to install meters would be prohibitively expensive.

[27] The Department of Energy's EIA is responsible for producing independent, unbiased research that helps the public, the federal government, industry, and the Congress better understand energy markets and promote sound policy making. EIA collects and analyzes data on the supply, consumption, and prices of oil and gas.

[28] EPA's estimates were based on publicly available oil and gas production data and information collected from industry participants in the Natural Gas STAR program, a nationwide, voluntary effort spearheaded by EPA aimed at reducing methane emissions from the oil and gas industry.

[29] Operators are required to report the sum of their vented gas and flared gas; they are not required to identify individual sources of lost gas.

[30] BLM's guidance was included in Notice to Lessees and Operators (NTL) 4A. Interior issues NTLs to clarify existing regulations. Operators need permission to vent or flare above routine amounts. In dealing with vented and flared gas, BLM's key guidance is in the form of an NTL. Offshore, BOEMRE uses regulations to guide operators in addressing vented and flared gas.

[31] Additional guidance from MRM explains how operators should submit data to the OGOR system, but does not provide detail on which sources to report, or on how they should be estimated.

[32] These limits generally allow operators to vent and flare gas required for routine well operations.

[33] See appendix I for more detail on how EPA developed its estimates.

[34] It is possible to estimate venting and flaring based on known emission rates of equipment type or production method. For example, if a pneumatic device is known to vent 10 cubic feet per hour, an operator would multiply that rate by the number of hours the piece of equipment operates each day.

[35] See appendix III for more details on the volumes and sources from the other four basins.

[36] See appendix I for more detail on the development of the estimates based on WRAP data.

[37] Offshore gas production in the Gulf of Mexico made up about 98 percent of total federal offshore natural gas production in 2008.

[38] Lease-use, or beneficial use, gas refers to natural gas that BLM and BOEMRE allow operators to use to power oil and gas production equipment on the lease. Emissions from pneumatic devices and glycol dehydrators would have been reported as lease-use gas, according to BOEMRE officials, and we determined it was not possible to extract these volumes from OGOR because they were combined with a number of other nonvented sources.

[39] The GOADS study included estimates of losses from natural gas compressors, although EPA's estimates were greater because of higher assumed losses from the compressor seals.

[40] BOEMRE officials told us that they did not draw conclusions from comparisons between 2005 GOADS and OGOR data because of the effect Hurricane Katrina and Rita had on offshore production in that year.

[41] Texas Commission for Environmental Quality (TCEQ), *Upstream Oil and Gas Storage Tank Project Flash Emissions Model Evaluation* (July 16, 2009).

[42] GAO-04-809.

[43] 75 Fed. Reg. 20291-20293 (April 19, 2010). The rule also requires vented gas and flared gas to be reported separately.

[44] For simplicity, EPA developed this estimate by focusing on the expansion potential of a subset of technologies considered to provide the largest emission reductions. The estimates may be conservative, however, because they did not incorporate reductions from a number of other potential venting and flaring opportunities catalogued by EPA's Natural Gas STAR program.

[45] Although there is likely some chance for similar reductions offshore, EPA did not estimate this amount.

[46] For more detail on carbon offsets see GAO, *Carbon Offsets: The U.S. Voluntary Market Is Growing, but Quality Assurance Poses Challenges for Market Participants*, GAO-08-1048 (Washington, D.C.: Aug. 29, 2008).

[47] For these calculations, we assumed average onshore royalty payments of 11.45 percent, the average onshore royalty rate in 2009. See appendix I for more details.

[48] Methane breaks down in the atmosphere more quickly than CO_2 and lasts an average of 12 years in the atmosphere. This accounts for its greater impact over the shorter time frame.

[49] This statement assumes that venting and flaring are reduced in proportional volumes.

[50] Overall, according to EPA's analysis, in addition to the total potential federal reductions of 50 Bcf, nonfederal wells could have added an additional 252 Bcf in reductions with more widespread use of venting and flaring reduction technologies.

[51] For onshore leases, *see* 43 C.F.R. § 3161.2. For offshore leases, *see* 30 C.F.R. § 250.106.

[52] BLM guidance is in the form of a Notice to Lessees and Operators (NTL). According to the NTL, the operator can vent or flare gas during operations such as clearing the drilling waste or removing liquid from the well for 24 hours without obtaining permission from BLM to vent gas. The operator may also flare or vent any gas vapors released from storage tanks or low pressure production vessels unless BLM determines that the recovery of the gas would be warranted. The vented or flared gas is considered to be "unavoidably lost."

[53] If an operator does not exceed these limits—which is almost always the case according to BLM staff—BLM does not consider the economic and technical viability of further reducing venting and flaring. BLM inspectors also note obvious signs of vented and flared gas during their inspections, which occur at least every 3 years, and try to verify that operators have permission for the release.

[54] In a previous report, we found that BOEMRE had not met these annual inspection goals. See GAO, *Oil and Gas Management: Interior's Oil and Gas Production Verification Efforts Do Not Provide Reasonable Assurance of Accurate Measurement of Production Volumes*, GAO-10-313 (Washington D.C.: Mar. 15, 2010).

[55] 43 C.F.R. § 3161.2.

[56] So far there has been only one rural oil and gas production basin, the Jonah-Pinedale basin in Wyoming, that is not meeting EPA standards for ground-level ozone.

[57] Applications include detailed information on plans for drilling and completing wells, such as the amounts and types of cement used, the construction materials, the methods for handling waste, the plans for surface reclamation, and multiple other subjects for BLM to consider. In addition, the operator submits a diagram of existing or proposed production facilities.

[58] 30 C.F.R. § 250.246.

[59] According to one BLM official we spoke with, an inspector may note whether or not operators use particular types of venting and flaring equipment, but the field office does not keep specific records about equipment use.

[60] One consulting firm we spoke with used infrared cameras to detect leaks on clients' offshore oil platforms and found, on average, 21 gas leaks per facility, totaling an estimated 127,000 cubic feet of gas per day.

[61] In addition to the Natural Gas STAR program, EPA's Office of Research and Development is developing specialized measurement approaches to remotely detect and quantify air emissions, including methane, from the oil and gas industry and other sources.

Natural Gas: Outlooks and Opportunities
Editor: Lucas N. Montauban

ISBN: 978-1-61324-132-5
© 2011 Nova Science Publishers, Inc.

Chapter 4

THE ALASKA NATURAL GAS PIPELINE: BACKGROUND, STATUS, AND ISSUES FOR CONGRESS[*]

Paul W. Parfomak

SUMMARY

Constructing a natural gas pipeline from Alaska's North Slope to the lower-48 states has been a government priority—periodically—for more than four decades. Beginning with the Alaska Natural Gas Transportation Act of 1976, Congress has repeatedly affirmed a national need for an Alaska gas pipeline. In remarks to the press President Obama has likewise described an Alaska gas pipeline as "a project of great potential ... as part of a comprehensive energy strategy."

Concerted efforts by Congress, the State of Alaska, and other stakeholders have resulted in new momentum to proceed with an Alaska gas pipeline project. A key step in advancing the pipeline project was the inclusion of $18 billion in federal loan guarantees for such a pipeline in the Alaska Natural Gas Pipeline Act of 2004 (P.L. 108-324). More recent milestones were the State of Alaska's 2008 award to TransCanada Corporation (TransCanada) of a license to build a natural gas pipeline from Prudhoe Bay to the lower-48 states and the concurrent announcement of a competing pipeline proposal, the Denali project, by BP and ConocoPhillips. Both projects plan to conduct open seasons in 2010 soliciting commitments for future gas shipments. If the open seasons yield sufficient interest from gas shippers, either the Denali or TransCanada project could file pipeline siting applications with the Federal Energy Regulatory Commission, Canada's National Energy Board, and other agencies. According to the Denali project's developers, the engineering, scheduling, and cost estimating work for the main pipeline and the associated gas treatment plant are continuing on schedule. TransCanada has been engaged in similar project activities on a comparable timeline.

Notwithstanding recent development progress, many potential obstacles to an Alaska gas pipeline remain at this time, especially the project's economics. A proposal in S.

---

[*] A version of this chapter was also published in Faces of *Alaska: Land Cover, Renewable Energy and Natural Gas Issues*, edited by Mark C. Monteith published by Nova Science Publishers, Inc. It was submitted for appropriate modifications in an effort to encourage wider dissemination of research.

1462 to increase the pipeline's federal loan guarantee to $30 billion, reported by the Senate Committee on Energy and Natural Resources on July 16, 2009, reflects continuing concerns about the project's economic viability due to low natural gas prices stemming from increased shale gas production. Nonetheless, the TransCanada and Denali developers indicate they are on track to begin construction of one of the proposed projects within the next few years. Proponents maintain that, if an Alaskan gas pipeline begins transporting gas to lower-48 markets by 2020 as anticipated, it would result in reduced U.S. energy prices, increased energy security, and lower U.S. emissions of carbon dioxide. They assert that the project would also create a significant number of jobs and support regional economic development. Like other infrastructure projects in wilderness areas, however, an Alaska gas pipeline would also involve significant environmental costs, many of which have yet to be determined.

Ultimately, the regional effects of any pipeline development on the Alaskan/Canadian environment must be weighed against its economic value, energy security value, and its global benefits in reducing carbon emissions from fossil fuels. To date, the judgment of Congress has favored construction of a pipeline—but ensuring that its public benefits continue to outweigh its costs will likely remain a key oversight challenge for the next decade.

INTRODUCTION

Over forty years ago, large natural gas reserves were discovered at Prudhoe Bay, Alaska. Since that time, Congress has been encouraging the development of those natural gas resources. Principal among its policies has been promoting the construction a natural gas pipeline from the Alaska North Slope to the lower-48 states. Beginning with the Alaska Natural Gas Transportation Act of 1976,[1] and continuing through the Alaska Natural Gas Pipeline Act of 2004 (P.L. 108-324), Congress has repeatedly affirmed a national need for an Alaska natural gas pipeline.[2] The presidential administrations of Jimmy Carter, Ronald Reagan, and George W. Bush also supported an Alaska gas pipeline project. The Obama administration continues that support. In remarks to the press during his first month in office, President Obama described an Alaska gas pipeline as "a project of great potential ... as part of a comprehensive energy strategy."[3]

While it has been on the drawing board for decades, on and off, interest in an Alaska natural gas pipeline has recently revived because of accelerated growth in U.S. natural gas demand, price volatility in the natural gas market, and the increased importation of liquefied natural gas (LNG) from overseas. Moreover, many analysts expect that a national policy of carbon dioxide control could further increase natural gas demand for electric power generation and, possibly, transportation fuel. These factors have led legislators to revisit the potential for an Alaska natural gas pipeline to meet future growth in U.S. demand for natural gas and to restart the process of Alaska gas pipeline development.[4] Important milestones in this activity were Alaska's August 2008 award to TransCanada Corporation (TransCanada) of a license to build a natural gas pipeline from Prudhoe Bay to the lower-48 states and the concurrent announcement of a competing pipeline proposal. To date, however, despite renewed support at both the federal and state levels, construction of an Alaska natural gas pipeline has not begun.

Issues for Congress

This report provides a brief review of the history of efforts to develop an Alaska natural gas pipeline, including project status, recent developments, and the current project outlook. It also discusses key policy issues related to the construction of the pipeline and its potential role in the context of U.S. energy and climate policy.[5] Specific issues of partricular interest to Congress include the implications for U.S. energy supplies and energy prices of an Alaska gas pipeline, and proposed legislation to raise the federal loan guarantee for the pipeline's construction. Other issues include an Alaska gas pipeline's environmental impacts, its physical security, and its relationship to the proposed Mackenzie Valley pipeline in Canada.

BACKGROUND

Arctic Alaska has substantial natural gas resources. The U.S. Geological Survey (USGS) estimates that conventional natural gas reserves on Alaska's North Slope potentially exceed 100 trillion cubic feet (Tcf), over four times the total annual gas consumption of the United States.[6] The agency's assessment of undiscovered conventional gas resources across the entire Arctic region concluded that over 1,600 Tcf of additional natural reserves gas remains to be found, with a significant share located under U.S. territory.[7] The USGS also estimates that the North Slope, specifically, may contain up to 158 Tcf of technically recoverable natural gas in the form of methane hydrates.[8] Although methane hydrate resources cannot yet be developed because there are no commercially viable methods to do so, future technologies may make such production economic. Taken together, these vast natural gas resources—both proven and anticipated—in Arctic Alaska have been the motivation for a natural gas pipeline to supply Alaskan natural gas to the lower-48 states.

In 1976, Congress passed the Alaska Natural Gas Transportation Act (ANGTA) to provide for sound decision-making on an Alaska Natural Gas Transportation System (ANGTS). The statute provided for congressional and presidential participation in the pipeline planning process and sought to expedite pipeline construction. The three main policy steps in the process were all completed by the end of 1977. First, the Federal Power Commission recommended (equally) two potential gas transportation options, both overland pipeline proposals along different routes through Alaska and Canada to the lower-48 states.[9] (A third, rejected, option was an overland pipeline across Alaska combined with an LNG export terminal in the port of Valdez.) President Carter then selected the route running along the Alaska Highway in preference to the northern route through the Mackenzie Delta, as shown in **Figure 1**. Congress approved the president's decision through a joint resolution on November 8, 1977.[10] This may have been the last time the project proceeded according to the original plan. A timeline including passage of ANGTS and other key events in the development of the Alaska gas pipeline project is presented in the **Appendix**.

Stalled Development

In the winter of 1977-1978, the United States experienced serious shortages of interstate natural gas supplies due to wellhead gas price controls imposed under a 1954 Supreme Court decision.[11] In response to these delivery problems, Congress passed the Natural Gas Policy Act of 1978 (NGPA), which effectively reversed the court's decision. Congress also passed the Powerplant and Industrial Fuel Use Act of 1978 (PIFUA), which restricted the construction of gas-fired power plants and the use of natural gas in large industrial boilers.[12] In the wake of these statutes and related policies, a natural gas oversupply developed, causing natural gas prices to fall sharply and persistently. At the same time, in an often repeated pattern for major U.S. energy supply projects, cost estimates for the Alaska gas pipeline increased. As a result of these factors, the desirability of the Alaska natural gas transportation system declined.[13] Commercial attention to the Alaska gas pipeline initiative essentially disappeared by the mid-1980s.

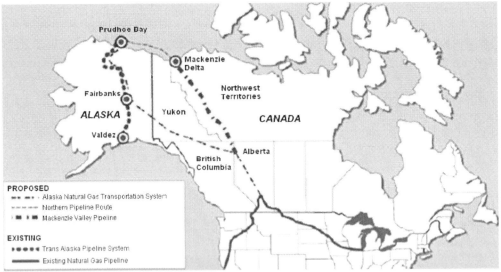

Source: Adapted from CRS Report RL31278, *Arctic National Wildlife Refuge: Background and Issues*, Figure 6, based on Energy Department maps.

Figure 1. Alaska Oil and Natural Gas Pipelines 2009.

Although the full Alaska gas pipeline system made little progress, the southernmost parts of the project did get developed during this period to transport growing natural gas supplies from the Western Canadian sedimentary basin. Producers from Alberta, along with Canadian authorities and U.S. and Canadian pipeline companies, cooperated to pre-build the downstream legs of the ANGTS. The western leg of the ANGTS was the Pacific Gas Transmission pipeline, which began service from Alberta to California in 1981. The eastern leg, the Northern Border Pipeline, went into service from Alberta to the Midwest in 1982 (**Figure 1**).

Marine Transport Options

In the 1980s, reacting to the lack of progress on the land-based pipeline system, the U.S. Maritime Administration authorized a study of marine options for transporting Alaskan natural gas as LNG aboard specialized tankers to determine whether they might offer U.S. shipbuilding opportunities. Such tankers had been transporting LNG to Japan from natural gas fields in south central Alaska (Cook Inlet) since 1969. The results of the study indicated roughly comparable economics for pipeline and LNG transport to the U.S. west coast. LNG sales to the Pacific Rim generally had greater economic potential, but were viewed as politically untenable because they would increase U.S. energy exports rather than increase domestic supplies.[14]

Renewed Interest in the Pipeline

Serious reconsideration of the construction of a natural gas pipeline from the Alaska North Slope began around 2000. This reconsideration was prompted, in large part, by tightening gas supplies to the lower-48 states and corresponding increases in natural gas prices and price volatility as shown in **Figure 2**.

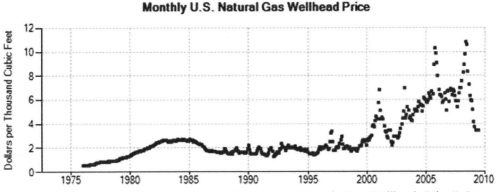

Source: U.S. Energy Information Administration, "U.S. Natural Gas Wellhead Price," Internet database, October 20, 2009, http://tonto.eia.doe.gov/dnav/ng/hist/n9190us3M.htm. Prices are in nominal dollars.

Figure 2. Growth and Volatility in U.S. Natural Gas Prices.

One important sign of renewed interest in an Alaska pipeline was the recommendation in the 2001 *National Energy Policy* report that the administration "expedite the construction of a pipeline to deliver natural gas to the lower 48 states."[15] Nearly concurrent with the release of this report, President George W. Bush issued Executive Order 13212, which set forth the administration's policy that executive departments and other federal agencies act to expedite projects that would increase the production, transmission, or conservation of energy.[16] To this end a federal interagency task force was established in 2001, co-chaired by the Departments of Energy and State, to identify any impediment to processing an Alaska gas pipeline permit application and to recommend ways to streamline the process.[17]

In 2004, Congress passed the Alaska Natural Gas Pipeline Act (P.L. 108-324, Div. C), a key step in advancing the pipeline project. This act originated in broader national energy legislation that was introduced in 2001, but advanced only slowly through Congress during 2002 and 2003. The Alaska provisions were eventually removed from the broader bill and enacted separately.[18] Among its key provisions, P.L. 108-324:

- clarified that, despite the passage of time since the earlier legislation, the Federal Energy Regulatory Commission (FERC) could still accept, review, and act upon applications for a new pipeline project under the Natural Gas Act or the Alaska Natural Gas Transportation Act;
- created an Office of the Federal Coordinator (OFC) for the issuance by other federal agencies of necessary pipeline permits;
- provided for a pipeline development loan guarantee of as much as $18 billion;
- established guidance for FERC to regulate the pipeline's capacity bidding process so that it would be available to parties beyond the three major North Slope producers—thereby promoting competition in Alaska North Slope natural gas development.[19]

The American Jobs Creation Act of 2004 (P.L. 108-357 § 706) provided a 7-year cost-of-investment recovery period for tax purposes (instead of the previously allowed 15-year period) and a designated economic life (class life) of 22 years for the Alaska natural gas pipeline. These provisions were intended by Congress to reduce the cost of capital for the project.[20] The act also provided a tax credit for a North Slope natural gas treatment plant required for the pipeline's operation. Broader, national energy legislation, which passed in 2005 (Energy Policy Act of 2005, P.L. 109-58), also addressed the Alaska natural gas pipeline. P.L. 109-58 required FERC to submit to Congress on a semi-annual basis progress reports about licensing and building the pipeline (§ 1810).[21] In June 2006, consistent with P.L. 108-324 and in accordance with E.O. 3212, fifteen federal agencies with regulatory and other responsibilities related to an Alaska natural gas pipeline signed a memorandum of understanding "to use best efforts to achieve early coordination and compliance with deadlines and procedures established by relevant agencies" in support of the pipeline's development.[22]

Alaska Gasline Inducement Act (AGIA)

In 2007, Alaska passed the Alaska Gasline Inducement Act (AGIA), seeking to encourage expedited construction of a natural gas pipeline from Alaska's North Slope to gas markets in Alaska and the lower-48 states.[23] AGIA requires a pipeline developer to meet certain requirements advancing the project in exchange for a license providing up to $500 million in matching funds, thereby reducing the developer's financial risks. Among other provisions, the statute requires a developer's commitment to: apply by a specific date for key regulatory approvals, to hold an open season for soliciting bids from North Slope gas producers for future shipments on the pipeline, and to expand the pipeline on rate and tariff terms that would allow gas producers to fully develop Alaska's North Slope natural gas resources.[24] The AGIA also establishes a competitive, public process for the evaluation and selection of pipeline proposals under specific criteria.

ALASKA GAS PIPELINE PROPOSALS

Developers have proposed seven major pipeline projects to develop North Slope natural gas reserves since 2007. Five of these proposals were filed under the AGIA process. Two have been proposed outside that process.

AGIA Proposals

In July 2007, the State of Alaska issued a Request for Applications (RFA) seeking proposals from any interested developer willing to meet the AGIA requirements. The state received competing applications from five developers:

- **TransCanada Pipelines Ltd. (TransCanada)** and Foothills Pipe Lines, Ltd., proposed a pipeline along a route similar to the ANGTS route for the originally approved project. The pipeline would connect with the existing TransCanada Alberta network at the Alberta hub. TransCanada would not construct a North Slope gas treatment plant to process wellhead gas for pipeline transport, although the company stated a willingness to develop such a plant if necessary. The original capacity of this pipeline would be between 4.5 to 5.0 billion cubic feet per day (Bcf/d). According to FERC staff, it would be expandable to 5.9 Bcf/d through the use of greater compression only, and, therefore, at relatively low added cost.[25] ExxonMobil, one of the three major Prudhoe Bay natural gas producers, joined with TransCanada to develop this project in June 2009.[26] The project anticipates an in-service date of 2018.
- **Alaska Gasline Port Authority (AGPA),** a municipal entity (including the City of Valdez, Fairbanks North Star Borough, and North Slope Borough), proposed a pipeline from the North Slope to Valdez, where the gas could be liquefied and shipped as LNG to Asia, the West Coast of the United States, Mexico, or Hawaii. The "All Alaska Gasline/LNG Project" would be constructed and operated by private contractors but publicly financed.
- **Alaska Natural Gasline Development Authority (ANGDA)** proposed a smaller capacity lateral pipeline to link from whatever major North Slope pipeline was selected to South Central Alaska (Anchorage and other locations), making up for declining production there.
- A **China Petroleum and Chemical Corporation (Sinopec)** subsidiary and Little Susitna Construction Company proposed a pipeline from the North Slope to Valdez in the south where the gas would be liquefied and exported to the Pacific Rim as LNG.
- **AEnergia LLC**, a startup company formed by persons with large project engineering experience, proposed a North Slope-to-Alberta pipeline jointly owned by the producers (74%), the State of Alaska (25%), and AEnergia (1%).[27]

Of these applications, former Alaska Governor Sarah Palin determined that only TransCanada's met the AGIA requirements and was complete enough to be evaluated. In

May 2008, the governor recommended the TransCanada proposal to the state legislature. The legislature accepted this recommendation and voted to award TransCanada the AGIA license in August 2008.[28] The Sinopec and AEnergia proposals subsequently were not pursued; the other three proposals are still active.

Other Proposals

In addition to the proposals filed under Alaska's AGIA process, developers have made two alternative proposals to construct North Slope gas pipelines or otherwise bring North Slope natural gas to market.

- **BP and Conoco Philips (Denali)**, two of the three major Prudhoe Bay natural gas producers, have jointly proposed a 4.0 Bcf/d pipeline project outside the AGIA process (and, thus, ineligible for the $500 million in state funds). The BP-Conoco Philips "Denali" Pipeline would follow a route similar to that of the TransCanada project to the Alberta hub. The Denali project, however, would include a gas treatment facility and would consider the option of new pipelines through Alberta and to the lower-48 states if deemed necessary.[29] The latest project schedule proposes initial pipeline service in 2018.[30]
- **Enstar Natural Gas Company** has proposed a 690-mile smaller capacity (0.5 Bcf/d) "bullet" pipeline from the North Slope to Southcentral Alaska as an alternative to the ANGDA lateral pipeline proposal. The Enstar pipeline would not depend upon the construction of a larger Alaska natural gas pipeline (either TransCanada's or Denali) to supply natural gas to in-state markets.[31]
- **Alaska Natural Resources To Liquids, LLC** is one of several "gas-to-liquids" (GTL) proposals that would convert North Slope natural gas to liquid hydrocarbon fuels (e.g., diesel and gasoline) and then transport them via a new pipeline or through the existing Trans Alaska Pipeline System (**Figure 1**) to Valdez for marine shipment out of state.[32] While not intended to deliver natural gas to the lower-48 states, such a project would develop Alaska's North Slope gas reserves as an alternative to a natural gas pipeline.

According to the Energy Information Administration, there are no "current" project sponsors actively promoting a GTL project on the North Slope.[33] Of the five Alaska natural gas pipeline projects that are still actively promoted, only the TransCanada, Denali, and AGPA proposals would transport natural gas out of Alaska. The other two proposals, ANGDA and Enstar, involve smaller projects focused exclusively on natural gas supplies to markets within Alaska. Because the latter two proposals are of more state interest than national interest, they are not discussed in the remainder of this report.

PROJECT ECONOMICS

An Alaska gas pipeline or pipeline/LNG terminal combination transporting natural gas from the North Slope to the lower-48 states would be, by some measures, the largest civilian construction project in the history of North America. When initially announced, the TransCanada and Denali pipeline developers estimated total project costs of $27 billion[34] and $30 billion,[35] respectively.

The Energy Information Administration currently assumes total capital costs for a TransCanada or Denali-type pipeline of approximately $28.8 billion and capital costs for a pipeline/LNG project of approximately $35.4 billion.[36] In light of these enormous capital requirements, policy makers and investors alike have paid close attention to key factors affecting the economic viability of the proposed projects, as well as their potential impact on regional and national economic development.

Natural Gas Price Expectations

Future prices of natural gas will have the greatest influence, by far, on the cost effectiveness of an Alaska natural gas pipeline. According to a 2008 study prepared for the Alaska legislature, the net present value of the TransCanada pipeline would be positive with long-term real natural gas prices above $5.00 per thousand cubic feet (Mcf) and potentially at lower prices.[37] A federal tax credit for the pipeline proposed in 2002 under H.R. 4 implied economic viability of the project at a gas price of $3.25/Mcf ($3.91/Mcf in 2009 dollars).[38] By comparison, the EIA's most recent analysis (which assumes an Alaska gas pipeline is constructed) projects lower-48 wellhead natural gas prices to rise from $4.64/Mcf in 2010 to $8.01/Mcf in 2030.[39] Based on such projections, assuming adequate gas production from North Slope fields, TransCanada expects its pipeline project to be cost-effective, generating $475 billion in total revenue over its operating life for distribution among gas producers and government stakeholders.[40] Assuming similar economics, the Denali project would also likely be cost effective.

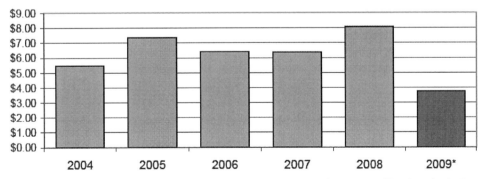

Source: U.S. Energy Information Administration, "U.S. Natural Gas Wellhead Price," Internet database, October 20, 2009, http://tonto.eia.doe.gov/dnav/ng/hist/n9190us3M.htm.
*2009 value is year-to-date through August.

Figure 3. Annual Average U.S. Wellhead Natural Gas Prices (Nominal dollars per thousand cubic feet).

Although the EIA's forecast of natural gas prices appears to support the construction of an Alaska gas pipeline, such forecasts are dynamic and can change unexpectedly due to structural changes in natural gas markets, changes in regulation, and general economic conditions. For example, when Congress passed the Alaska Natural Gas Pipeline Act of 2004, the average annual price for U.S. wellhead natural gas was $5.46/Mcf and rising (**Figure 3**). Four years later, in 2008, the average annual gas price increased to $8.07/Mcf. This high price level, however, spurred unexpected growth in the production of "unconventional" natural gas from shale deposits in the lower 48-states.[41] The combination of new shale gas supplies and reduced growth in gas demand due to the recent U.S. economic downturn subsequently caused natural gas prices to plummet. For the first eight months of 2009, wellhead natural gas prices averaged less than $4.00/Mcf. While demand for natural gas will likely rebound as the U.S. economy resumes its growth, independent gas producers assure strong lower-48 domestic natural gas supplies for the foreseeable future—continuing to put downward pressure on gas prices.[42] In such a price environment, there may be ongoing concerns as to the cost-effectiveness of an Alaska gas pipeline; long-term pipeline economics will be a constantly moving target.[43] Nonetheless, Alaska state officials reportedly have stated that, notwithstanding unprecedented shale gas development across North America, the outlook for the pipeline "is extremely healthy from an economic standpoint."[44] The state's Alaska Pipeline Project Report, released in October 2009, likewise states that "the temporary weakness of North American natural gas demand is not a determining factor in establishing the economic viability of an Alaska natural gas pipeline."[45]

Construction Costs and Cost Uncertainties

TransCanada and Denali have estimated similar costs for their proposed projects—between $28 and $30 billion. Some analysts, however, suggest that these estimates are conservative and that project costs could reach $40 billion.[46] Infrastructure analysts point to numerous examples of cost overruns for other recent multi-billion dollar, decade-long transportation projects, such as Boston's "Big Dig" Central Artery/Tunnel Project, the English Channel Tunnel, and the Oresund Bridge from Denmark to Sweden.[47] They conclude that "large construction cost escalations in transport infrastructure projects are common and exist across different project types, different continents, and different historical periods."[48] While the TransCanada and Denali pipeline developers claim extensive experience managing the construction of major pipeline projects within budget, the scale of an Alaska gas pipeline would be unprecedented, and the risk of cost overruns may be significant. Projected costs for a similar but smaller Arctic natural gas pipeline proposal, the Mackenzie Valley Pipeline (further discussed below), which is also still in development, reportedly doubled between 2005 and 2007.[49]

Economic Development

The Alaska natural gas pipeline project has important implications for Alaska's economy. By one estimate commissioned by the state in 2006, the net present value of state and local government earnings from the project, including associated North Slope natural gas production, could exceed $29 billion over its lifetime.[50] New natural gas royalty revenue could offset future declines in ongoing oil royalties from mature North Slope oil fields, which have accounted for 85% of the state's unrestricted general fund, on average, since the Trans Alaska Pipeline System began operation.[51] Although state oil revenues set a record in FY2009 due to high global oil prices, oil production in Alaska has actually declined 63% from its peak in 1988.[52]

Development of an Alaska natural gas pipeline to the lower-48 states would also create many new jobs. The Alaska-commissioned analysis concluded in 2006 that the pipeline

> increases the state's labor force needs by an average of 18,000 direct, indirect and induced workers per year during construction. The project also creates a sustained impact of about 26,000 jobs per year after construction from both pipeline operations and jobs generated by state and local spending of project-related oil and gas revenue.[53]

Throughout the nation, numerous additional jobs would be created in support of, or resulting from, the pipeline's construction. According to one estimate cited in 2004 by Senator Lisa Murkowski, up to 1.1 million jobs nationwide could be created—directly and indirectly—by the pipeline, although details of this analysis are not publicly available.[54] A more recent analysis of public infrastructure spending reported that 21,888 total jobs are created for every $1 billion in new investment in natural gas projects.[55] Based on this ratio, a $28 billion Alaska gas pipeline would create approximately 600,000 new jobs, including jobs in Canada. These jobs include those directly involved in constructing the new project, supplying the project (e.g., line pipe, construction equipment), and induced by the general spending of new pipeline-related employees. Projections of new jobs are highly uncertain, however, particularly for a project as unique as an Alaska gas pipeline, so the actual number of jobs to be created remains somewhat speculative.

In addition to job creation, the presence of a major interstate pipeline from the North Slope to the lower-48 could change the economics of many local energy markets within Alaska. For example, both TransCanada and Denali have agreed to provide up to five natural gas delivery points within Alaska, which could allow local communities near the pipeline right-of-way to provide new natural gas service. Delivery to Anchorage and other Southcentral Alaska markets could also offset declining natural gas supplies from gas fields at Cook Inlet. New or sustained supplies of natural gas to these communities would be an important factor in their economic health.

CURRENT PROJECT STATUS

Because the pipeline/LNG proposals for transporting natural gas out of Alaska would involve interstate commerce, they are subject to federal siting approval by FERC under the Natural Gas Act of 1938.[56] Additional state and local approvals for aspects of the project would be required, as well as approvals from Canadian federal, territorial, provincial, and First Nation agencies for the sections of the project running through Canada.

The Denali and TransCanada pipelines, only one of which could be constructed, are proceeding along similar tracks. Both have initiated FERC's pre-filing process, whereby prospective pipeline developers engage with stakeholders—including state, local, and other federal agencies—before filing an application for a Natural Gas Act 7(c) pipeline certificate. Although Denali began the pre-filing process in June 2008, and TransCanada in April 2009, both projects plan to conduct open seasons in 2010 soliciting commitments for future gas shipments. If the open seasons yield sufficient interest from gas shippers, either the Denali or TransCanada project could file pipeline siting applications with FERC, Canada's National Energy Board, and other agencies. According to the Denali's developers, the engineering, scheduling, and cost estimating work for the main pipeline and the associated gas treatment plant are continuing on schedule.[57] TransCanada has been engaged in similar project activities on a comparable timeline.[58] Based on these development schedules, the EIA, Alaska officials, and other stakeholders anticipate that an Alaska natural gas pipeline to the lower-48 states will be in service by approximately 2020.[59]

The TransCanada and Denali pipeline projects are competing and mutually exclusive proposals, each backed by different North Slope producers. To reduce project costs and improve the likelihood of success, some key policy makers have been encouraging the various developers to come together on a single pipeline project. For example, Alaska Governor Sean Parnell has stated, "Four major, international companies all working on an Alaska gas line. To me, that's good news. Now the challenge is to get them working on the same project."[60] To date, the projects are continuing independently, but producer cooperation on a single pipeline is a possibility.

The LNG Option

The AGPA proposal to build a pipeline from the North Slope to Valdez, and then ship out natural gas on LNG tankers, has not advanced to the detailed planning and project development stage. However, TransCanada has committed to including the option of transporting natural gas for the AGPA's proposed LNG project within its open season for its mainline pipeline from the North Slope to Alberta.[61] According to remarks to Alaska legislators in June 2009, TransCanada has seen "serious interest" among potential shippers for an LNG alternative to its all-land pipeline proposal.[62] Interest in advancing the LNG project has further increased since September 2009, when the AGPA's general counsel, Bill Walker, declared his candidacy for the Alaska governorship with the AGPA gas pipeline/LNG project as the primary issue in his campaign.[63]

POLICY ISSUES FOR CONGRESS

Congress's last major action to promote an Alaska gas pipeline was passing the Alaska Natural Gas Pipeline Act in 2004. Under that federal statutory framework, as well as provisions in Alaska state law, both the TransCanada and Denali pipeline projects appear to be proceeding on schedule. Since 2004, Congress has been monitoring progress on pipeline development, with only a few active legislative proposals directly addressing these projects (further discussed below). As Congress continues to oversee Alaska gas pipeline activities, several policy issues may attract particular attention. The most important of these is continued uncertainty about whether an Alaska gas pipeline will ultimately be constructed—and, if not, the broader implications for U.S. energy supplies and energy prices. Other important issues are Canada's development of a Mackenzie Valley Pipeline, the federal loan guarantee for the Alaskan pipeline's construction, the environmental impact of the project, and its physical vulnerability to acts of terrorism.

Alaska Gas Pipeline Development Uncertainty

Many policy makers and developers have high expectations for the contribution of an Alaska natural gas pipeline to U.S. energy supplies. By 2030, the EIA projects an Alaska pipeline carrying 1.99 trillion cubic feet (Tcf) of natural gas annually, which is about 8.5% of total projected U.S. gas supplies.[64] But given the scale, investment requirements, and 10-year time frame required to construct such a pipeline, there continues to be uncertainty regarding its completion. As discussed above, the project faces specific risks deriving from gas prices and pipeline construction costs—but it also faces significant risks involving natural gas production costs, environmental impacts, interest rates, tax policies, and regulatory uncertainty. If, through some combination of these factors, the pipeline's development stalls (again), the anticipated Alaskan natural gas supplies may not become available to lower-48 markets, which may have a ripple effect on the U.S. energy sector.

- **Potentially higher natural gas prices.** Without Alaskan gas supplies, U.S. gas markets would have fewer supply options, and, therefore, might face higher prices. According to the State of Alaska, EIA, and other studies, an Alaska pipeline is expected to be economically viable supplying 1.99 Tcf of natural gas at a market price of $4.00/Mcf to $5.00/Mcf. In the anticipated gas market environment, such a large quantity of relatively low cost gas would help to moderate growth and volatility of natural gas prices in the lower-48 states. This might be especially true if the alternatives for incremental U.S. gas supplies are costlier unconventional gas supplies from North America, or imported LNG—which has a history of price volatility in the international market.
- **Increased LNG imports.** If LNG requirements increase, then U.S. dependence on overseas energy supplies must also increase, counter to a principal energy policy objective of President Obama and many members of Congress. In its current forecast, the EIA projects LNG imports to account for 0.81 Tcf in 2030, or about 3.5% of U.S. natural gas supplies.[65] Absent an Alaska gas pipeline, the United States will have sufficient LNG import capacity in place to double or triple LNG imports if global gas prices warrant it.[66]
- **Carbon control challenges.** Because burning natural gas produces only half as much carbon dioxide as burning coal, and also less carbon dioxide than vehicular fuels like gasoline, many view natural gas as the preferred fuel in the near-term for power plants and motor vehicles under a national policy of carbon control.[67] Not having Alaskan natural gas supplies might, therefore, make it more difficult or costly to reach U.S. carbon reduction targets by substituting natural gas for other fossil fuels in power generation or motor vehicles.[68]

Canada's Mackenzie Valley Pipeline

Because large sections of a proposed Alaska natural gas pipeline to the lower-48 states would pass through Canada, the involvement of Canadian agencies is essential for the pipeline's success. Canada has cooperated with the United States for decades on a variety of general matters related to the development of an Alaska natural gas pipeline (e.g., the pre-build segments in the early 1980s). However, Canada has been pursuing its own Artic natural gas projects as well. The Canadian project of greatest importance, which could affect an Alaska natural gas pipeline project, is the development of a Mackenzie Valley pipeline.

The Mackenzie River Delta region in the Canadian Arctic contains an estimated 40 Tcf of natural gas.[69] The Canadian government has been interested in developing a pipeline to transport this natural gas to North American markets since it was first discovered in the 1970s in a process that has, in many ways, paralleled the Alaska gas pipeline's development. The main proposal has been a new pipeline from the delta through the Mackenzie River Valley to the existing natural gas pipeline network in Alberta, as illustrated earlier in **Figure 1**. Initially, developers also proposed transporting North Slope gas through a spur pipeline connecting to the proposed Mackenzie pipeline, but the Alaskan connection was rejected by a Canadian government inquiry report in 1974, three years before President Carter also rejected this option.[70] The current configuration of a stand-alone pipeline to Alberta is similar in design to the Alaska gas pipeline proposals, albeit smaller in capacity (1.2 Bcfd) and lower in cost ($15 billion) with a projected in-service date of 2016.[71]

Although there has been significant activity to advance the development of the Mackenzie Valley Pipeline, the project has been controversial due to environmental impacts and escalating costs.[72] According to press reports in October 2009, the Canadian government appeared to be reconsidering the level of financial support it planned to provide the project due to "concerns about the project's price tag."[73] The pipeline's developers maintain that it remains viable, however, and the project continues to receive government support as they proceed with project groundwork and meeting regulatory requirements.[74] Canada's National Energy Board plans to hear final arguments regarding the pipeline's siting application in April 2010, after which it will render a decision as to whether the project will be allowed to proceed.[75]

Because the Mackenzie Valley pipeline would commercialize a major new source of North American natural gas, and would draw on a limited pool of construction resources and materials available for such a project, it has been viewed by some as a direct competitor to an Alaska gas pipeline.[76] For example, the demand for steel and pipe for a Mackenzie Valley project would be significant, and it is not clear that there is adequate, large diameter pipe production capacity in the entire world to supply both Alaska and Mackenzie Valley projects at the same time. Other analysts anticipate that large increases in North American natural gas demand over the next decade will support development of both pipeline projects, and that sufficient construction materials will become available when they are needed. They view the Alaska and Mackenzie Valley pipeline projects as complementary rather than competitive.[77] While the future of natural gas prices and the availability of construction commodities is difficult to predict, because of their physical similarities and links to the same infrastructure, the fates of the Mackenzie and Alaska gas pipelines may remain intertwined.

Federal Loan Guarantee

As mentioned above, to reduce the financial risks associated with the development of an Alaska natural gas pipeline, the Alaska Natural Gas Pipeline Act of 2004 authorizes the Department of Energy (DOE) to issue up to $18 billion of project loan guarantees for up to 80% of its capital costs and a term of 30 years. The Consolidated Appropriations Act of 2005 (P.L. 108-447) extended the availability of loan guarantees to developers of an LNG project, such as that proposed by AGPA, that would transport natural gas from Southcentral Alaska to West Coast States.[78] DOE solicited comments and information from the public in 2005 in advance of a possible rulemaking concerning the loan guarantee provisions, but has not issued such rules.[79]

The agency reportedly is waiting for the developers to submit more complete project plans before proceeding with specific loan program requirements.[80]

Federal loan guarantees for an Alaska natural gas pipeline have been controversial because they are viewed by some stakeholders, especially supporters of Mackenzie Pipeline development, as an unfair form of subsidy.[81] They also expose the federal government to a significant portion of the financial risk associated with the project. The Congressional Budget Office estimated in 2003 that the loan guarantee under P.L. 108-324 would involve a 10% subsidy and cost $2 billion over the 2010-2013 period, although the agency acknowledged that this estimate was "quite uncertain."[82] Some in Congress are calling for greater federal loan guarantees for an Alaska gas pipeline. Responding to the recent fall in natural gas prices and tighter credit markets in the wake of the U.S. banking crisis, Senators Bingaman and Murkowski have included provisions in the American Clean Energy Leadership Act of 2009 (S. 1462) that would raise the limit from $18 billion to $30 billion (§ 353), among other provisions. The Senate Committee on Energy and Natural Resources reported S. 1462 on July 16, 2009.

Environmental Impacts

The addition of significant natural gas resources from Alaska's North Slope to the lower-48 states' fuel supplies is considered by many policy makers to be environmentally beneficial because natural gas produces much lower atmospheric emissions than other fossil fuels (although still far more than renewables, nuclear power, and conservation). Nonetheless the immensity of the roughly 1,750-mile construction project and associated development of North Slope natural gas fields has caused concern about potential environmental effects on land and wildlife. These concerns were heightened by the 2006 shutdown of North Slope oil fields following the discovery of severe corrosion and a small spill from one of BP's Prudhoe Bay oil pipelines.[83] Broader environmental impacts from existing North Slope oil development and the associated TAPS pipeline have been significant and are well-documented.[84]

Congress has addressed the direct environmental effects of Alaska gas pipeline development by designating FERC as the lead federal agency in preparing an environmental impact statement under the National Environmental Policy Act of 1969 (P.L. 91-190). Under this statute, pipeline permit applicants must prepare certain environmental reports to aid FERC in its preparation of the environmental impact statement (18 C.F.R. § 380.3). The

Environmental Protection Agency, as a supporting agency, often assists in the review of the environmental reports and the issuance of the environmental impact statements. In some cases, however, FERC's environmental reviews of major natural gas projects have been challenged in the courts.[85] Given the scale and uniqueness of an Alaska gas pipeline project, future environmental controversies may arise.

One specific environmental issue that faces Alaska gas pipeline development is the potential use of North Slope gas in Alberta's oil (tar) sands industry. North Slope natural gas, if developed, may be used to fuel oil sand operations in northern Alberta. These operations require large quantities of steam to extract crude oil from regional bitumen deposits. Some stakeholders object to the potential use of natural gas for crude oil production on the grounds that it, in their view, consumes a "clean" fuel to produce a "dirty" one, and because oil sands projects can have significant local environmental impacts.[86] The debate over the environmental impacts and potential energy security benefits of Canadian oil sands production is beyond the scope of this report. Nonetheless, given that the primary rationale for federal support of an Alaska natural gas pipeline is to increase lower-48 natural gas supplies, there is the potential for misunderstanding if substantial volumes of North Slope gas are instead diverted to the Canadian oil sands industry as Alaska natural gas begins flowing. This issue may require policy attention.

Terrorism Risks and Infrastructure Concentration

An Alaska natural gas pipeline would be integral to U.S. energy supply and have vital links to other critical infrastructure, such as power plants, airports, and military bases. While an efficient and fundamentally safe means of transport, such a pipeline would carry volatile material with the potential to cause public injury and environmental damage—either accidentally or intentionally. For these reasons, similar pipelines have been favored targets of terrorists in North America and overseas. Since September 11, 2001, federal warnings about Al Qaeda have mentioned pipelines specifically as potential terror targets in the United States.[87] Congress has responded with substantial initiatives to protect U.S. pipelines from such attacks.[88] Nonetheless, due to its great length and passage through remote areas, an Alaska gas pipeline could still be vulnerable to vandalism and terrorist attack.

Concerns about the security of an Alaska gas pipeline are especially significant because of a recent history of physical attacks on existing pipelines in Alaska and Canada. For example, between October 2008 and July 2009, natural gas pipelines in British Columbia, Canada were bombed six times by unknown perpetrators.[89] In 1999, Vancouver police arrested a man planning to blow up the Trans Alaska Pipeline System (TAPS) for personal profit in oil futures.[90] In 2001, a vandal's attack on TAPS with a high-powered rifle forced a two-day shutdown and caused extensive economic and ecological damage.[91] In January 2006, federal authorities acknowledged the discovery of a detailed posting on a website purportedly linked to Al Qaeda that reportedly encouraged attacks on U.S. pipelines, especially TAPS, using weapons or hidden explosives.[92] In November 2007 a U.S. citizen was convicted of trying to conspire with Al Qaeda to attack TAPS and a major natural gas pipeline in the eastern United States.[93] To date, there have been no known Al Qaeda attacks on TAPS or other U.S. pipelines, but such attacks remain a possibility.

Construction of an Alaska gas pipeline directly alongside TAPS also raises concerns because it increases the concentration of critical U.S. energy infrastructure in the same narrow geographic corridor. When infrastructure is physically concentrated in such a limited geographic area it may be particularly vulnerable to geographic hazards such as natural disasters and certain kinds of terrorist attacks. Whereas a typical geographic disruption is often expected to affect infrastructure in proportion to the size of an affected region, a disruption of concentrated infrastructure could have greatly disproportionate—and national—effects. In 2005, Hurricanes Katrina and Rita demonstrated this kind of geographic impact by disrupting a substantial part of the national U.S. energy and chemical sectors, both heavily concentrated in the Gulf of Mexico. In 2008, Hurricanes Gustav and Ike caused similar disruptions, renewing concerns about geographic vulnerability. Adding 8.5% of U.S. natural gas supplies to a right-of-way that currently delivers nearly 17% of the nation's domestic oil production would be a significant increase in the physical concentration of U.S. energy flows.[94] As development of an Alaska natural gas pipeline continues, physical threats and required security measures may be an important policy consideration.

CONCLUSION

Constructing an Alaska natural gas pipeline from the North Slope to the lower-48 states has been a government priority—on and off and on again—for more than four decades. Concerted efforts by Congress, the State of Alaska, and other stakeholders have resulted in new momentum to proceed with the project. Many potential obstacles remain at this time, especially the project's economics. The proposal in S. 1462 to increase the pipeline's federal loan guarantee reflects continuing concerns about the project's economic viability. Nonetheless, the developers project the start of construction within the next few years. If the pipeline begins transporting gas to lower-48 markets as anticipated by 2020, it could have an impact on U.S energy prices, energy security, and U.S. emissions of carbon dioxide. The project would also create a significant number of jobs and support regional economic development. Like other infrastructure projects in wilderness areas, however, an Alaska gas pipeline would also involve significant environmental costs, many of which have yet to be determined.

Ultimately, the regional effects of the pipeline's development on the Alaskan/Canadian environment must be weighed against its economic value, energy security value, and its global benefits in reducing carbon emissions from fossil fuels. To date, the judgment of Congress has favored the pipeline—but ensuring that its public benefits continue to outweigh its costs will likely remain a key oversight challenge for the next decade. If the balance tips the other way— either in the eyes of developers or the federal government—and the Alaska gas pipeline is not constructed, Congress may need to redouble its support of other energy initiatives that could fill the substantial expectations unmet by this project.

APPENDIX

Table 1. Key Events in Alaska Natural Gas Pipeline Development

1968	Prudhoe Bay oil and gas discovered
1969	United States begins export of LNG to Japan from south central Alaska (Cook Inlet)
1976	Alaska Natural Gas Transportation Act (ANGTA) passed, P.L. 94-586
1977	Presidential Decision and FPC Report to Congress on ANGTS
1977	FERC (successor to FPC) issues conditional certificate for pipeline
1978	Trans-Alaskan Pipeline System (TAPS) oil pipeline into service
1981	"Western leg" of Alaska gas pipeline (Pacific Gas Transmission) into service
1982	"Eastern leg" (Northern Border Pipeline) into service
1983	Maritime Administration study of marine alternatives to ANGTS pipeline released
2001	Alaska Natural Gas Interagency Task Force established
2004	Alaska Natural Gas Pipeline Act passed, P.L. 108-324, Division C
2006	New governor announces Alaska Gasline Inducement Act (AGIA) initiative
2007	Five proposals submitted for AGIA consideration
2008	Governor determines one AGIA proposal meets AGIA criteria
2008	Conoco Phillips and BP announce the Denali Project as an alternative to an AGIA project
2008	Alaska legislature approves governor's AGIA recommendation and it becomes law
2009	American Clean Energy Leadership Act of 2009 (S. 1462) proposed to raise loan guarantee

End Notes

[1] 15 U.S.C. § 719 et seq.

[2] For example, see P.L. 108-324 § 103(b)(2)(A) "a public need exists to construct and operate the proposed Alaska natural gas transportation project."

[3] Erika Bolstad, "Obama Calls Alaska Gas Pipeline Promising," *Anchorage Daily News*, February 11, 2009.

[4] U.S. Energy Information Administration, *Natural Gas Monthly*, August 2009, Table 2. Full initial pipeline capacity is 4.5 billion cubic feet per day.

[5] Portions of this report previously appeared in CRS Report RL34671, *The Alaska Natural Gas Pipeline: Status and Current Policy Issues*, by William F. Hederman.

[6] David W. Houseknecht, U.S. Geological Survey, *Conventional Natural Gas Resource Potential, Alaska North Slope*, Open File Report 2004-1440, December 13, 2004; U.S. Energy Information Administration, *Annual Energy Outlook 2009*, DOE/EIA-0383(2009), March 2009, p. 109.

[7] U.S. Geological Survey, *USGS Arctic Oil and Gas Report*, Fact Sheet, July 2008.

[8] U.S. Geological Survey, *Gas Hydrate Resource Assessment: North Slope, Alaska*, Fact Sheet, October 2008

[9] The Federal Power Commission (FPC) was the predecessor of the Federal Energy Regulatory Commission (FERC).

[10] H.R. J. Res. 621, P.L. 95-158, 91 Stat. 1268, 95th Cong., 1st Sess., 1977.

[11] Phillips Petroleum Co. v. Wisconsin, 347 U.S. 672

[12] P.L. 95-62 and P.L. 95-620.

[13] See discussion in D. Fried and W. Hederman, "The Benefits of an Alaska Natural Gas Pipeline," *The Energy Journal*, Vol. 2, No. 1, January 1981, p. 22.

[14] W. F. Hederman, "A Review of Marine Systems Use in Developing Alaska Natural Gas," SPE 11294, *SPE Hydrocarbon Economics and Evaluation Symposium*, March 2, 1983, Dallas, TX.

[15] National Energy Policy Development Group, *National Energy Policy*, Office of the Vice President, May 16, 2001, p. 7-18.

[16] E.O. 13212, "Actions to Expedite Energy-Related Projects," 66 F.R. 28357, signed by President Bush on May 18, 2001.

[17] U.S. Department of Energy, National Energy Technology Laboratory, *Delivering Alaskan North Slop Gas To Market*, June 2004, p. 8, http://www.netl.doe.gov/publications/factsheets/policy/Policy003.pdf. Other task force members included the Department of the Interior (Bureau of Land Management and Minerals Management Service), the Department of Transportation (Office of Pipeline Safety), and the Federal Energy Regulatory Commission.

[18] State of Alaska, Department of Revenue, *The Alaska Natural Gas Pipeline Act of 2004*, undated, p. 1, http://www.revenue.state.ak.us/gasline/Annex%201.pdf

[19] The FERC issued a final rule on the open season matter on February 9, 2005 (FERC Order No. 2005).

[20] U.S. Congress, Joint Committee on Taxation, *General Explanation Of Tax Legislation Enacted In The 108th Congress*, Joint committee print, JCS-5-05, May 2005, p. 330.

[21] P.L. 109-58, 119 Stat. 594(2005), 42 U.S.C. Section 15801 *et seq.*

[22] U.S. Dept. of Agriculture, et al. *,Memorandum of Understanding Related to an Alaska Natural Gas Transportation Project*, June 2006, p. 3, http://www.usace.army.mil/CECW/ Documents/cecwo/reg/mou/Alaska_MOU.PDF.

[23] AS 43.90 et seq.

[24] State of Alaska, "Frequently Asked Questions Regarding the Alaska Gasline Inducement Act ("AGIA") and the AGIA Process," November 8, 2008, p. 1, http://www.gov.state.ak.us/agia/agia_faqs_11808.pdf.

[25] Federal Energy Regulatory Commission, *Sixth Report to Congress on Progress Made in Licensing and Constructing the Alaska Gas Pipeline*, August 29, 2008, p. 8.

[26] TransCanada Corp., "TransCanada and ExxonMobil to Work Together on Alaska Pipeline Project," Press release, June 11, 2009.

[27] Federal Energy Regulatory Commission, *Fifth Report to Congress on Progress Made in Licensing and Constructing the Alaska Gas Pipeline*, February 19, 2008, pp. 2-3.

[28] Alaska HB 3001, August 1, 2008.

[29] BP and ConocoPhillips, "BP and ConocoPhillips Join on the Alaska Gas Pipeline," Joint press release, April 8, 2008.

[30] BP and ConocoPhillips, "Denali - The Alaska Gas Pipeline Project," Web page, October 15, 2009, http://www.denalipipeline.com/overview.php.

[31] Fairbanks Economic Development Corp., *In-State Gas Pipeline Supply Options Study*, February 5, 2009, p. 39.

[32] Alaska Natural Resources To Liquids, LLC, "A Legacy Decision for Alaska," Presentation to the Alaska State Legislature, Joint Legislative Budget and Audit Committee, June 20, 2008; Tim Bradner, "Natural Gas: Exxon, BP, Independent All Planning Gas-To-Liquids Projects," *Alaska Journal of Commerce*, April 16, 2000.

[33] U.S. Energy Information Administration, *Annual Energy Outlook 2009*, DOE/EIA-0383(2009), March 2009, p. 41.

[34] TransCanada Corp., *Application for License: Alaska Gasline Inducement Act*, November 30, 2007, p. 2.5-2. 2007 dollars.

[35] Wesley Loy, "BP, Conoco Join Forces to Pursue Gas Pipeline," *Anchorage Daily News*, April 9, 2008. 2008 dollars.

[36] U.S. Energy Information Administration, March 2009, p. 39. Costs are adjusted to 2009 dollars. A gas-to-liquids option is assumed to have capital costs of approximately $60 billion.

[37] Black & Veatch, *Net Present Value (NPV) Analysis*, Presentation to the Alaska State Legislature, Joint Legislative Budget and Audit Committee, June 18, 2008, p. 18, http://lba.legis.state.ak.us/proposals/doclog/2008-06- 18black_veatch_npv_spec_sess_jnt_comm.pdf.

[38] H.R. 4, Energy Policy Act of 2002, Engrossed Amendment as Agreed to by Senate, Section 2503. The provision was not enacted.

[39] U.S. Energy Information Administration, *An Updated Annual Energy Outlook 2009 Reference Case*, SR/OIAF/2009-03, April 2009, p. 44. Prices are in 2007 dollars.

[40] Leslie Ferron-Jones, TransCanada Corp., *Projects Impacting Western Markets*, Presentation to the California Energy Commission, 2009 Integrated Energy Policy Report Workshop, Sacramento, CA, May 14, 2009, p. 8, http://www.energy.ca.gov/2009_energypolicy/documents/2009-05-14_workshop/presentations/13_TransCanada_Ferron_Jones.pdf.

[41] U.S. Energy Information Administration, *Is U.S. Natural Gas Production Increasing?*, Energy Brief, June 11, 2008, http://tonto.eia.doe.gov/energy_in_brief/natural_gas_production.cfm.

[42] See, for example, Navigant Consulting, Inc., *The State of North American Natural Gas Supply: Review of Groundbreaking Findings*, Presentation to the California Energy Commission, 2009 Integrated Energy Policy Report Workshop, Sacramento, CA, May 14, 2009, http://www.energy.ca.gov/ 2009_energypolicy/documents/2009-05-14_workshop/presentations/05_Navigant_Pickering_Smead_Natural_Gas_Supply_Assessment.pdf.

[43] See, for example: Elizabeth Bluemink, "Gas Pipeline Project Faces 'Fierce' Competition," *Anchorage Daily News*, September 15, 2009.

[44] Rena Delbridge, "State Sticks by Pipeline Deal," *Alaska Dispatch*, October 26, 2009.

[45] Alaska Dept. of Revenue and Dept. of Natural Resource, *Alaska Pipeline Project Report: Licensed under the Alaska Gasline Inducement Act (AGIA)*, October 31, 2009. p. 21.

[46] Tim Bradner, "Wobbly Market May Crimp Gas Line," *Alaska Journal of Commerce*, February 2, 2009.

[47] A. E. Barinov, "Systemic and Political Factors Affecting Cost Overrun in the World Economy's Large Investment Projects," *Studies on Russian Economic Development*, Vol. 18, No. 6, 2007, pp. 650–658.

[48] Bent Flyvbjerg, Mette K. Skamris Holm And Søren L. Buhl, "What Causes Cost Overrun in Transport Infrastructure Projects?," *Transport Reviews*, Vol. 24, No. 1, January 2004, p. 16.

[49] "Former Canadian Official: State Gas Line Faces Numerous Hurdles," *Associated Press*, April 15, 2008.

[50] Information Insights, Inc., *Economic, Fiscal and Workforce Impacts of Alaska Natural Gas Projects*, Prepared for Alaska Department of Revenue, May 10, 2006, p. 8. Adjusted to 2009 dollars.

[51] Resource Development Council for Alaska, Inc., "Alaska's Oil and Gas Industry," Web page, November 11, 2009, http://www.akrdc.org/issues/oilgas/overview.html.

[52] Ibid.

[53] Information Insights, Inc., May 10, 2006, p. 82.

[54] U.S. Congress, House Committee on Energy and Commerce, Subcommittee on Energy and Air Quality, May 6, 2004, p. 10.

[55] Political Economy Research Institute, How Infrastructure Investments Support the U.S. Economy: Employment, Productivity and Growth, January, 2009, p. 25.

[56] Section 7 of the Natural Gas Act authorizes FERC to issue certificates of "public convenience and necessity" for "the construction or extension of any facilities ... for the transportation in interstate commerce of natural gas" (15 U.S.C. § 717f).

[57] Steven Findlay, Vice President, Denali—The Alaska Gas Pipeline, LLC., *September Monthly Status Report*, FERC Docket PF08-26, Accession No. 20091007-5035, October 7, 2009, p. 2.

[58] TransCanada Alaska Company LLC., *August Monthly Status Report*, FERC Docket PF09-11, Accession No. 20090915-5025, September 15, 2009; Tim Bradner, "Gas Pipeline Plans On Schedule But Bumps Lie Ahead," *Alaska Journal of Commerce*, September 18, 2009.

[59] U.S. Energy Information Administration, March 2009, p. 77.

[60] Stefan Milkowski, "Parnell Chamber Talk Focuses on Energy," *Petroleum News*, Vol. 14, No. 40, October 4, 2009.

[61] Federal Energy Regulatory Commission, *Eighth Report to Congress on Progress Made in Licensing and Constructing the Alaska Gas Pipeline*, August 26, 2009, p. 6.

[62] Tony Palmer, Vice President, TransCanada Corp., Testimony before the Alaska House Resources Committee, June 23, 2009.

[63] Bill Walker, All-Alaska Governor, Web page, October 15, 2009, http://www.billwalkerforgovernor.com.

[64] U.S. Energy Information Administration, April 2009, pp. 42, 44.

[65] U.S. Energy Information Administration, April 2009, pp. 42, 44.

[66] Federal Energy Regulatory Commission, "Existing LNG Terminals," and "Approved North American LNG Import Terminals," Web pages, October 27, 2009, http://www.ferc.gov/industries/lng.asp.

[67] See, for example: John D. Podesta and Timothy E. Wirth, *Natural Gas: A Bridge Fuel for the 21st Century*, Center for American Progress, August 10, 2009, http://www.americanprogress.org/issues/ 2009/08/pdf/naturalgasmemo.pdf; Rena Delbridge, "Federal Energy Commissioner Checks on Alaska Pipeline Projects," *Alaska Dispatch*, September 25, 2009.

[68] For further discussion and analysis of U.S. greenhouse gas policy, see CRS Report RL34513, *Climate Change: Current Issues and Policy Tools*, by Jane A. Leggett.

[69] U.S. Geological Survey, "Assessment of Undiscovered Oil and Gas Resources of the Mackenzie Delta Province, North America, 2004," Fact Sheet 2006-3002, March, 2006.

[70] Thomas R. Berger, Commissioner, *Northern Frontier, Northern Homeland: The Report of the Mackenzie Valley Pipeline Inquiry, Volume One*, Ministry of Supply and Services Canada, April 15, 1977, p. xiii.

[71] Federal Energy Regulatory Commission, *Seventh Report to Congress on Progress Made in Licensing and Constructing the Alaska Gas Pipeline*, February 20, 2009, p. 6.

[72] Claudia Cattaneo, "Little Hope in the Pipeline," *Financial Post*, July 1, 2009.

[73] John Ivison and Carrie Tait, "Pipeline Dream in Peril," *National Post*, October 27, 2009.

[74] Jeffrey Jones, "Imperial Oil CEO 'Dismayed' by Mackenzie Report," *Reuters*, November 3, 2009.

[75] National Energy Board Canada, "National Energy Board Announces Next Steps In Mackenzie Gas Project Hearing," Press release, October 7, 2009.

[76] See, for example: James Irwin, "Alaska Pipeline Advance Could Threaten Mackenzie Valley Pipeline," *Natural Gas Week*, March, 7, 2005.

[77] U.S. Congress, House Committee on Energy and Commerce, Subcommittee on Energy and Air Quality, *Alaska Natural Gas Pipeline Status Report*, 108th Cong., 2nd sess., May 6, 2004, 108-82 (Washington: GPO, 2004), p. 15.

[78] Title I, Division J § 114.

[79] U.S. Department of Energy, "Alaska Natural Gas Pipeline Loan Guarantee," 70 *Federal Register* 30707-30708, May 27, 2005.

[80] Federal Energy Regulatory Commission, August 26, 2009, p. 8.

[81] Stephen Barlas, "Energy Bill Likely to Bless Alaska NG Pipeline," *Pipeline & Gas Journal*, November 2002; "U.S. Loan Guarantees for Alaska Pipeline Worry Mackenzie Line Supporters," *CBC News*, May 27, 2009.

[82] Congressional Budget Office, *S. 14 Energy Policy Act of 2003, As Introduced on April 30, 2003,* Cost Estimate, May 7, 2003, p. 7, http://www.cbo.gov/doc.cfm?index=4206&type=0.

[83] For further background see archived CRS Report RL33629, *BP Alaska North Slope Pipeline Shutdowns: Regulatory Policy Issues*, by Paul W. Parfomak.

[84] National Research Council, *Cumulative Environmental Effects of Oil and Gas Activities on Alaska's North Slope*, Committee on the Cumulative Environmental Effects of Oil and Gas Activities on Alaska's North Slope, National Academies Press, Washington, DC, 2003.

[85] For example, see Jim Springhetti, "State Asks Court to Toss Bradwood Site's Approval," *The Oregonian*, January 26, 2009.

[86] IHS Cambridge Energy Research Associates, *Growth in the Canadian Oil Sands: Finding the New Balance*, Cambridge, MA, 2009, pp. III-1-III-20; Andrew Nikiforuk, *Dirty Oil: How the Tar Sands are Fueling the Global Climate Crisis*, Greenpeace, September 2009.

[87] "Already Hard at Work on Security, Pipelines Told of Terrorist Threat," *Inside FERC*, McGraw-Hill Companies, January 3, 2002.

[88] For additional analysis of pipeline security issues, see CRS Report RL33347, *Pipeline Safety and Security: Federal Programs*, by Paul W. Parfomak.

[89] Elise Stolte, "EnCana Puts Record $1M on Bomber's Head," *Edmonton Journal*, July 31, 2009.

[90] D. S. Cloud, "A Former Green Beret's Plot to Make Millions Through Terrorism," *Ottawa Citizen*, December 24, 1999, p. E15.

[91] Y. Rosen, "Alaska Critics Take Potshots at Line Security," *Houston Chronicle*, February 17, 2002.

[92] W. Loy, "Web Post Urges Jihadists to Attack Alaska Pipeline," *Anchorage Daily News*, January 19, 2006.

[93] U.S. Attorney's Office, Middle District of Pennsylvania, "Man Convicted of Attempting to Provide Material Support to Al-Qaeda Sentenced to 30 Years' Imprisonment," Press release, November 6, 2007; A. Lubrano and J. Shiffman, "Pa. Man Accused of Terrorist Plot," *Philadelphia Inquirer*, February 12, 2006, p. A1.

[94] Alyeska Pipeline Service Co., Internet page, Anchorage, AK, November 2, 2009, http://www.alyeska-pipe.com/about.html.

Natural Gas: Outlooks and Opportunities
Editor: Lucas N. Montauban
ISBN: 978-1-61324-132-5
© 2011 Nova Science Publishers, Inc.

Chapter 5

LIQUEFIED NATURAL GAS (LNG) IMPORT TERMINALS: SITING, SAFETY, AND REGULATION[*]

Paul W. Parfomak and Adam Vann

SUMMARY

Liquefied natural gas (LNG) is a hazardous fuel shipped in large tankers to U.S. ports from overseas. While LNG has historically made up a small part of U.S. natural gas supplies, rising price volatility, and the possibility of domestic shortages have significantly increased LNG demand. To meet this demand, energy companies have proposed new LNG import terminals throughout the coastal United States. Many of these terminals would be built onshore near populated areas.

The Federal Energy Regulatory Commission (FERC) grants federal approval for the siting of new onshore LNG facilities under the Natural Gas Act of 1938 and the Energy Policy Act of 2005 (P.L. 109-58). This approval process incorporates minimum safety standards for LNG established by the Department of Transportation. Although LNG has had a record of relative safety for the last 45 years, and no LNG tanker or land-based facility has been attacked by terrorists, proposals for new LNG terminal facilities have generated considerable public concern. Some community groups and governments officials fear that LNG terminals may expose nearby residents to unacceptable hazards. Ongoing public concern about LNG safety has focused congressional attention on the exclusivity of FERC's LNG siting authority, proposals for a regional LNG siting process, the lack of "remote" siting requirements in FERC regulations, state permitting requirements under the Clean Water Act and the Coastal Zone Management Act, terrorism attractiveness of LNG, the adequacy of Coast Guard security resources, and other issues.

LNG terminals directly affect the safety of communities in the states and congressional districts where they are sited, and may influence energy costs nationwide. Faced with an uncertain national need for greater LNG imports and persistent public concerns about LNG hazards, some in Congress have proposed changes to safety provisions in federal LNG siting regulation. Legislation proposed in the 110th Congress addressed Coast Guard LNG resources, FERC's exclusive siting authority, state

---

[*] This is an edited, reformatted and augmented version of a Congressional Research Services publication, dated December 14, 2009.

concurrence of federal LNG siting decisions, and agency coordination under the Coastal Zone Management Act, among other proposals. Provisions in the Coast Guard Authorization Act of 2010 (H.R. 3619), passed by the House on October 23, 2009, would require additional waterway suitability notification requirements in LNG siting reviews by FERC (Sec. 1117). The Maritime Hazardous Cargo Security Act (S. 1385), introduced by Senator Lautenberg and three co-sponsors on June 25, 2009, would require a national study to identify measures to improve the security of maritime transportation of liquefied natural gas (Sec. 6).

If Congress concludes that new LNG terminals as currently regulated will pose an unacceptable risk to public safety, Congress may consider additional LNG safety-related legislation, or may exercise its oversight authority in other ways to influence LNG terminal siting approval. Alternatively, Congress may consider other changes in U.S. energy policy legislation to reduce the nation's demand for natural gas or increase supplies of North American natural gas and, thus, the need for new LNG infrastructure.

INTRODUCTION

Liquefied natural gas (LNG) historically has played a minor role in U.S. energy markets, but in reaction to rising natural gas prices, price volatility, and the possibility of domestic shortages, demand for LNG imports has increased significantly in recent years. To meet anticipated growth in LNG demand, new onshore and offshore LNG import terminals have been constructed or approved in United States coastal regions. More have been proposed. Because LNG (like other fossil fuels) is a hazardous[1] liquid transported and stored in enormous quantities—often near populated areas—concerns exist about the federal government's role in addressing LNG safety in the terminal siting process. In addition, various energy policy proposals could impact the need for new LNG terminals by encouraging the development of alternative U.S. energy supplies and promoting conservation and efficiency.

This report provides an overview of recent industry development of new LNG import terminals. The report summarizes LNG hazards and the industry's safety record. It discusses federal laws and regulations related to LNG terminal siting with a focus on the authorities of key federal agencies and safety provisions in the permitting of onshore facilities. The report reviews controversial safety issues in recent LNG siting proceedings, such as safety zones, marine hazards, hazard modeling, and remote siting. The report outlines policy issues related to LNG terminal safety, including the Federal Energy Regulatory Commission's (FERC's) LNG siting authority, regional LNG siting, "remote" siting requirements in federal regulations, state permitting requirements, terrorism, and other issues.

Issues Facing Congress

LNG terminals directly affect the safety of communities in the states and congressional districts where they are sited, and may influence energy costs nationwide. Faced with an uncertain national need for greater LNG imports and persistent public concerns about LNG hazards, some in Congress have proposed changes to safety provisions in federal LNG siting regulation. Legislation proposed in the 110th Congress addressed Coast Guard LNG resources, FERC's exclusive siting authority, state concurrence of federal LNG siting decisions, and agency coordination under the Coastal Zone Management Act, among other proposals.[2] If Congress concludes that new LNG terminals as currently regulated will pose an unacceptable risk to public safety, Congress may consider additional LNG safety-related legislation, or may exercise its oversight authority in other ways to influence LNG terminal siting approval. Alternatively, Congress may consider other changes in U.S. energy policy legislation to reduce the nation's demand for natural gas or increase supplies of North American natural gas and, thus, the need for new LNG infrastructure.

Scope and Limitations

This report focuses broadly on industry and federal activities related to safety in LNG import terminal siting. For a more specific discussion of LNG security, see CRS Report RL32073, *Liquefied Natural Gas (LNG) Infrastructure Security: Issues for Congress*, by Paul W. Parfomak.

This report also deals primarily with those parts of LNG terminals which transfer, store, and process LNG prior to injection to natural gas pipelines for transmission off site. For more discussion of general natural gas or pipeline hazards, see CRS Report RL33347, *Pipeline Safety and Security: Federal Programs*, by Paul W. Parfomak. Also, this report discusses mostly onshore facilities and near-shore shipping, since they pose the greatest public hazards. Offshore LNG terminal siting regulations are summarized in the Appendix.

BACKGROUND

What Is LNG and Where Does It Come From?

When natural gas is cooled to temperatures below minus 260° F it condenses into liquefied natural gas, or LNG. As a liquid, natural gas occupies only 1/600th the volume of its gaseous state, so it is stored more effectively in a limited space and is more readily transported. A single tanker ship, for example, can carry huge quantities of LNG—enough to supply a single day's energy needs of over 10 million homes. When LNG is warmed it "regasifies" and can be used for the same purposes as conventional natural gas such as heating, cooking, and power generation.

In 2009, LNG imports to the United States originated in Trinidad (54%), Egypt (34%), Norway (8%), and Nigeria (4%).[3] In recent years, some LNG shipments have also come from Algeria, Qatar, Equatorial Guinea, Malaysia, Oman, Australia, and other countries.[4] Brunei, Indonesia, Libya, and the United Arab Emirates also export LNG, and may be U.S. suppliers in the future. In addition to importing LNG to the lower 48 states, the United States exports Alaskan LNG to Japan.

Expectations for U.S. LNG Import Growth

The United States has used LNG commercially since the 1940s. Initially, LNG facilities stored domestically produced natural gas to supplement pipeline supplies during times of high gas demand. In the 1970s, LNG imports began to supplement domestic production. Primarily because of low domestic gas prices, LNG imports stayed relatively small—accounting for only 1% of total U.S. gas consumption as late as 2002.[5] In countries with limited domestic gas supplies, however, LNG imports grew dramatically over the same period. Japan, for example, imported 97% of its natural gas supply as LNG in 2002, more than 11 times as much LNG as the United States.[6] South Korea, France, Spain, and Taiwan also became heavy LNG importers.

Natural gas demand growth accelerated in the United States from the mid-1980s through 2000 due to environmental concerns about other energy sources, widespread building of natural gas-fired electricity generation, and low natural gas prices. Domestic gas supplies have not always kept up with growth in demand, however, so prices have become volatile. At the same time, international LNG costs have fallen since the 1970s because of increased supplies and more efficient production and transportation, making LNG more competitive with domestic natural gas.

In 2003 testimony before Congress, the Federal Reserve Chairman called for a sharp increase in LNG imports to help avert a potential barrier to U.S. economic growth. According to the Chairman's testimony: "... high gas prices projected in the American distant futures market have made us a potential very large importer.... Access to world natural gas supplies will require a major expansion of LNG terminal import capacity."[7] Likewise, FERC Commissioner Suedeen Kelly told industry representatives in 2006 that,"while LNG has made a marginal contribution to gas supply over the last 30 years, it is poised to make a major contribution in the future."[8] Because burning natural gas produces only half as much carbon dioxide as burning coal, and also less carbon dioxide than vehicular fuels like gasoline, some also anticipate natural gas demand to grow as the preferred fuel in the near-term for power plants and motor vehicles under a national policy of carbon control.[9] Recent increases in U.S. natural gas production from domestic shale deposits have complicated projections about LNG markets. Nonetheless, many analysts expect continued growth in the U.S. LNG imports over the long term.[10]

Proposed LNG Import Terminals in the United States

LNG tankers unload their cargo at dedicated marine terminals which store and regasify the LNG for distribution to domestic markets. Onshore terminals consist of docks, LNG handling equipment, storage tanks, and interconnections to regional gas transmission pipelines and electric power plants. Offshore terminals regasify and pump the LNG directly into offshore natural gas pipelines or may store LNG for later injection into offshore pipelines.

There are seven active onshore LNG import terminals in the United States: Everett, Massachusetts; Lake Charles, Louisiana; Cove Point, Maryland; Elba Island, Georgia; Peñuelas, Puerto Rico; Freeport, Texas; and Sabine Pass, Louisiana. There are two active offshore import terminals, one located in the Gulf of Mexico and a second near Boston,

Massachusetts. (There is also one export terminal in Kenai, Alaska.) In addition to these active terminals, some 25 LNG terminal proposals have been approved by regulators across North America to serve the U.S. market **Figure 1**. A number of these proposals have been withdrawn, however, due to siting problems, financing problems, or other reasons. Developers have proposed another 8 U.S. terminals prior to filing formal siting applications.[11]

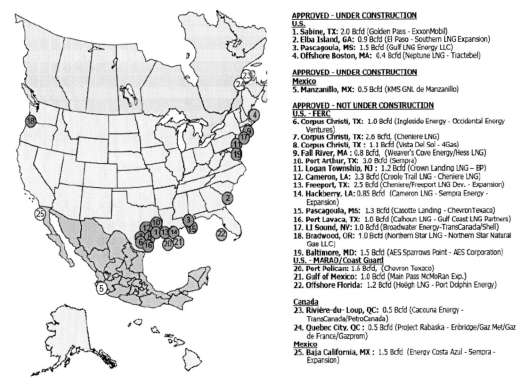

Source: Federal Energy Regulatory Commission (FERC), "Approved North American LNG Import Terminals," updated October 27, 2009. http://ferc.gov/industries/lng/indus-act/terminals/lng-approved.pdf.

Figure 1. Approved LNG Terminals in North America.

POTENTIAL SAFETY HAZARDS FROM LNG TERMINALS

The safety hazards associated with LNG terminals have been debated for decades. A 1944 accident at one of the nation's first LNG facilities killed 128 people and initiated public fears about LNG hazards which persist today.[12] Technology improvements and standards since the 1940s have made LNG facilities much safer, but serious hazards remain since LNG is inherently volatile and is usually shipped and stored in large quantities. A 2004 accident at Algeria's Skikda LNG terminal, which killed or injured over 100 workers, added to the ongoing controversy over LNG facility safety.[13]

Physical Hazards of LNG

Natural gas is combustible, so an uncontrolled release of LNG poses a hazard of fire or, in confined spaces, explosion. LNG also poses hazards because it is so cold. The likelihood and severity of catastrophic LNG events have been the subject of controversy. While questions remain about the credible impacts of specific LNG hazards, there appears to be consensus as to what the most serious hazards are.

Pool Fires

If LNG spills near an ignition source, evaporating gas will burn above the LNG pool.[14] The resulting "pool fire" would spread as the LNG pool expanded away from its source and continued evaporating. A pool fire is intense, burning far more hotly and rapidly than oil or gasoline fires.[15] It cannot be extinguished—all the LNG must be consumed before it goes out. Because an LNG pool fire is so hot, its thermal radiation may injure people and damage property a considerable distance from the fire itself. Many experts agree that a large pool fire, especially on water, is the most serious LNG hazard.[16]

Flammable Vapor Clouds

If LNG spills but does not immediately ignite, the evaporating natural gas will form a vapor cloud that may drift some distance from the spill site. If the cloud subsequently encounters an ignition source, those portions of the cloud with a combustible gas-air concentration will burn. Because only a fraction of such a cloud would have a combustible gas-air concentration, the cloud would not likely ignite all at once, but the fire could still cause considerable damage.[17] An LNG vapor cloud fire would gradually burn its way back to the LNG spill where the vapors originated and would continue to burn as a pool fire.[18]

Other Safety Hazards

LNG spilled on water could (theoretically) regasify almost instantly in a "flameless explosion," although an Idaho National Engineering Laboratory report concluded that "transitions caused by mixing of LNG and water are not violent."[19] LNG vapor clouds are not toxic, but they could cause asphyxiation by displacing breathable air.[20] Such clouds may begin near the ground (or water) when they are still very cold, but rise in air as they warm, diminishing the threat to people. Due to its extremely low temperature, LNG could injure people or damage equipment through direct contact.[21] Such contact would likely be limited, however, as a major spill would likely result in a more serious fire. The environmental damage associated with an LNG spill would be confined to fire and freezing impacts near the spill since LNG dissipates completely and leaves no residue.[22]

Terrorism Hazards

LNG tankers and land-based facilities could be vulnerable to terrorism. Tankers might be physically attacked in a variety of ways to release their cargo—or commandeered for use as weapons against coastal targets. LNG terminal facilities might also be physically attacked with explosives or through other means. Some LNG facilities may also be indirectly disrupted by "cyber-attacks" or attacks on regional electricity grids and communications networks which could in turn affect dependent LNG control and safety systems.[23] The potential attractiveness of LNG infrastructure to terrorists as a target is discussed later in this report.

Safety Record of LNG

The LNG tanker industry claims a record of relative safety over the last 50 years; since international LNG shipping began in 1959, tankers have carried over 45,000 LNG cargoes and traveled over 128 million miles without a serious accident at sea or in port.[24] LNG tankers have experienced groundings and collisions during this period, but none has resulted in a major spill.[25] The LNG marine safety record is partly due to the double-hulled design of LNG tankers. This design makes them more robust and less prone to accidental spills than old single-hulled oil and fuel tankers like the *Exxon Valdez*, which caused a major Alaskan oil spill after grounding in 1989.[26] LNG tankers also carry radar, global positioning systems, automatic distress systems and beacons to signal if they are in trouble. Cargo safety systems include instruments that can shut operations if they deviate from normal as well as gas and fire detection systems.[27]

The safety record of onshore LNG terminals is more mixed. There are more than 40 LNG terminals (and more than 150 other LNG storage facilities) worldwide. Since 1944, there have been approximately 13 serious accidents at these facilities directly related to LNG. Two of these accidents caused single fatalities of facility workers—one in Algeria in 1977, and another at Cove Point, Maryland, in 1979. On January 19, 2004, a fire at the LNG processing facility in Skikda, Algeria killed an estimated 27 workers, injured 74 others, destroyed a processing plant, and damaged a marine berth. (It did not, however, damage a second processing plant or three large LNG storage tanks also located at the terminal, nor did the accident injure the rest of the 12,000 workers at the complex.)[28] It was considered the worst petrochemical plant fire in Algeria in over 40 years.[29] According to press reports, the accident resulted from poor maintenance rather than a facility design flaw.[30] Another three accidents at worldwide LNG plants since 1944 have also caused fatalities, but these were construction or maintenance accidents in which LNG was not present.

LNG Hazard Models

Since the terror attacks of September 11, 2001, a number of technical studies have been commissioned to reevaluate the safety hazards of LNG terminals and associated shipping. The most widely cited of these studies are listed in Table 1. These studies have caused controversy because some reach differing conclusions about the potential public hazard of LNG terminal accidents or terror attacks. Consequently, some fear that LNG hazards may be misrepresented by government agencies, or that certain LNG hazards may simply not be understood well enough to support a terminal siting approval.[31]

Hazard analyses for LNG terminals and shipping depend heavily upon computer models to approximate the effects of hypothetical accidents. Federal siting standards specifically require computer modeling of thermal radiation and flammable vapor cloud exclusion zones (49 C.F.R. §§ 193.2057, 2059).[32] Such models are necessary because there have been no major LNG incidents of the type envisioned in LNG safety research and because historical LNG experiments have been limited in scale and scope. But LNG hazards models simulate complex physical phenomena and are inherently uncertain, relying on calculations and input assumptions about which fair-minded analysts may legitimately disagree. Even small differences in an LNG hazard model have led to significantly different conclusions. Referring to previous LNG safety zone studies, for example, FERC noted in 2003 that "distances have been estimated to range from 1,400 feet to more than 4,000 feet for [hazardous] thermal radiation."[33]

Table 1. Recent LNG Hazard Studies

Author	Sponsor	Subject
Lloyd's Register of Shipping[a]	Distrigas (Tractebel)	Focused models of possible terror attacks on LNG ships serving Everett
Quest Consultants Inc.[b]	DOE (lead), FERC, DOT	Models catastrophic breach of an LNG ship tank
James Fay (MIT)[c]	Fair Play for Harpswell	Models fire and vapor hazards of proposed Harpswell LNG terminal
Tobin & Associates[d]	City of Vallejo	Reviews general safety of proposed Mare Island LNG terminal
Lehr and Simecek-Beatty[e]	NOAA staff	Compares hypothetical LNG and fuel oil fires on water
Det Norske Veritas[f]	LNG Industry Companies	Models LNG maximum credible failures
ABSG Consulting[g]	FERC (lead), DOT, USCG	Reviews consequence assessment methods for LNG tanker incidents
Sandia National Laboratories[h]	DOE	Two reports examine effect of large-scale LNG spills on water (additional studies underway)

Source: Congressional Research Service.

[a] Waryas, Edward. Lloyd's Register Americas, Inc. "Major Disaster Planning: Understanding and Managing Your Risk." Fourth National Harbor Safety Committee Conference. Galveston, TX. March 4, 2002. Summary excerpts are in this presentation.

[b] Juckett, Don. U.S. Department of Energy (DOE). "Properties of LNG." LNG Workshop. Solomons, MD. February 12, 2002. A Quest study summary is in this presentation.

[c] Fay, James A. "Public Safety Issues at the Proposed Harpswell LNG Terminal." FairPlay for Harpswell. Harpswell, ME. November 5, 2003.

[d] Tobin & Assoc. "Liquefied Natural Gas in Vallejo: Heath and Safety Issues." Report by the LNG Safety Committee of the Disaster Council, Vallejo, CA. January 16, 2003.

[e] Lehr, W. and Simecek-Beatty, D. "Comparison of Hypothetical LNG and Fuel Oil Fires on Water." *Journal of Hazardous Materials*. v. 107. 2004. pp. 3-9.

[f] Pitblado, R.M., J. Baik, G. J. Hughes, C. Ferro, and S. J. Shaw., Det Norske Veritas. "Consequences of LNG Marine Incidents." Presented at the Center for Chemical Process Safety (CCPS) Conference. Orlando, FL. June 29-July 1, 2004. Available at http://www.dnv.com/press/dnvcompletes studyonlngmarinereleases.asp.

[g] ABSG Consulting. *Consequence Assessment Methods for Incidents Involving Releases from Liquefied Natural Gas Carriers*. GEMS 1288209. Prepared for the Federal Energy Regulatory Commission under contract FERC04C40196. May 13, 2004.

[h] Sandia National Laboratories (SNL). Breach and Safety Analysis of Spills Over Water from Large Liquefied Natural Gas Carriers. SAND2008-3153. Albuquerque, NM. May 2008; Sandia National Laboratories (SNL). Guidance on Risk Analysis and Safety Implications of a Large Liquefied Natural Gas (LNG) Spill Over Water. SAND2004-6258. Albuquerque, NM. December 2004.

The LNG hazard studies in **Table 1** have been sponsored by a range of stakeholders and have been performed by individuals with various kinds of expertise. It is beyond the scope of this report to make detailed comparisons of the methodologies and findings of these studies and FERC analysis. Furthermore, each of the available studies (or its application) appears to have significant limitations, or has been questioned by critics. For example, the ABSG Consulting study released by FERC in May 2004, which reviewed existing LNG hazard models, concluded that

- No release models are available that take into account the true structure of an LNG carrier;
- No pool spread models are available that account for wave action or currents; and
- Relatively few experimental data are available for validation of models involving LNG spills on water, and there are no data available for spills as large as the spills considered in this study.[34]

The 2004 Sandia National Laboratories study similarly reported that "there are limitations in existing data and current modeling capabilities for analyzing LNG spills over water."[35] Nonetheless, the Sandia report concluded that "existing [analytic] tools ... can be used to identify and mitigate hazards to protect both public safety and property."[36]

Uncertainty related to LNG hazard modeling continues. A December 2006 study using yet another LNG computer model of a large LNG fire states that "current generation models that are being used to calculate the radiant heat ... from the fire are found to be overly conservative."[37] In February 2007, the Government Accountability Office (GAO) issued a report comparing six recent unclassified studies (including studies in **Table 1**) of the consequences of LNG spills. The GAO report concluded that[38]

> Because there have been no large-scale LNG spills or spill experiments, past studies have developed modeling assumptions based on small-scale spill data. While there is general agreement on the types of effects from an LNG spill, the results of these models have created what appears to be conflicting assessments of the specific consequences of an LNG spill, creating uncertainty for regulators and the public.

Following the GAO report, Members of Congress expressed renewed concern about the uncertainty associated with LNG hazard analysis.[39]

Hazards vs. Risks

In reviewing the various LNG hazard studies, it is important to be clear about the distinction between *hazards* and *risks*. Although theoretical models may try to quantify the effects of "worst-case" hazards, evaluating the risks associated with those hazards requires an estimate of the probability that they will occur. Some argue that a significant *hazard* that is nonetheless highly unlikely does not represent an unacceptable *risk* to the public. In this view, worst-case hazard studies alone do not provide a sufficient basis for evaluating public safety. Unfortunately, few LNG safety studies comprehensively and convincingly address the probability of catastrophic accidents or attacks actually occurring.[40] In part, this shortcoming arises from a lack of historical LNG incidents and detailed terrorist threat information on which to base such probabilities. Faced with this analytic uncertainty, decision makers are forced to draw the best information they can get and rely upon their own best judgment to reach conclusions about LNG safety.

LNG Terminal Safety in Perspective

Other Hazardous Materials

LNG terminals and tankers have a high profile because of extensive media coverage, although there are few of them relative to all the hazardous chemical plants and ships currently operating near U.S. cities. According to the U.S. Environmental Protection Agency (EPA), for example, more than 500 toxic chemical facilities operate in "urban" areas at which worst-case accidents could affect 100,000 or more people.[41] These include chlorine plants in city water systems and ammonia tanks in agricultural fertilizer production. There are also oil refineries and liquefied petroleum gas (e.g., propane, butane) terminals operating in U.S. ports that pose safety hazards similar to those of LNG. Based on the most recent data available from the U.S. Office of Hazardous Materials Safety, there are over 100,000 annual U.S. shipments of hazardous marine cargo such as ammonia, crude oil, liquefied petroleum gases, and other volatile chemicals.[42] Many of these cargoes pose a hazard similar to LNG and pass through the same harbors serving existing or proposed LNG terminals.

Civil and Criminal Liability

One reason LNG tanker and terminal operators seek to ensure public safety is to avoid civil and criminal liability from an LNG accident; there are no special provisions in U.S. law protecting the fossil fuel industry from such liability. As a result of the 1989 *Exxon Valdez* oil spill, for example, Exxon has been required to pay over $500 million in criminal and civil settlements.[43] In January 2003, the Justice Department announced over $100 million in civil and criminal penalties against Olympic Pipeline and Shell Pipeline resolving claims from a fatal pipeline fire in Bellingham, Washington in 1999.[44] In March 2003, emphasizing the environmental aspects of homeland security, the U.S. Attorney General reportedly announced a crackdown on companies failing to protect against possible terrorist attacks on storage tanks, transportation networks, industrial plants, and pipelines.[45] In 2002, federal safety regulators proposed a $220,000 fine against the Distrigas LNG terminal in Everett, Massachusetts, reportedly for security training violations.[46] Notwithstanding these actions, some observers are skeptical that government scrutiny will ensure LNG infrastructure safety.

Even if no federal or state regulations are violated, LNG companies could still face civil liability for personal injury or wrongful death in the event of an accident. In the Bellingham case, the pipeline owner and associated defendants reportedly agreed to pay a $75 million settlement to the families of two children killed in the accident.[47] In 2002, El Paso Corporation settled wrongful death and personal injury lawsuits stemming from a natural gas pipeline explosion near Carlsbad, New Mexico, which killed 12 campers.[48] Although the terms of those settlements were not disclosed, two additional lawsuits sought a total of $171 million in damages. The impact of these lawsuits on the company's business is unclear, however; El Paso's June 2003 quarterly financial report stated that "our costs and legal exposure ... will be fully covered by insurance."[49]

REGULATION OF ONSHORE LNG SITING

The Department of Transportation (DOT) and FERC are the federal agencies primarily responsible for the regulation of onshore LNG facilities. Although federal statutes do not explicitly designate the relative jurisdiction of DOT and FERC, the agencies have clarified their roles through interagency agreement. These roles and their relation to other authorities are summarized below.

Department of Transportation

The DOT sets safety standards for onshore LNG facilities. The DOT's authority originally stemmed from the Natural Gas Pipeline Safety Act of 1968 (P.L. 90-481) and the Hazardous Liquids Pipeline Safety Act of 1979 (P.L. 96-129). These acts were subsequently combined and recodified as the Pipeline Safety Act of 1994 (P.L. 102-508). The acts were further amended by the Pipeline Safety Improvement Act of 2002 (P.L. 107-355) and the Pipeline Safety Improvement Act of 2006 (P.L. 109-468). Under the resulting statutory scheme, DOT is charged with issuing minimum safety standards for the siting, design, construction, and operation of LNG facilities. It does not approve or deny specific siting proposals, because that authority is vested with FERC, as discussed below.

The Pipeline Safety Act, as amended, includes the following provisions concerning LNG facility siting (49 U.S.C. § 60103):

> The Secretary of Transportation shall prescribe minimum safety standards for deciding on the location of a new liquefied natural gas pipeline facility. In prescribing a standard, the Secretary shall consider the—
>
> 1) kind and use of the facility;
> 2) existing and projected population and demographic characteristics of the location;
> 3) existing and proposed land use near the location;
> 4) natural physical aspects of the location;
> 5) medical, law enforcement, and fire prevention capabilities near the location that can cope with a risk caused by the facility; and
> 6) need to encourage remote siting.

General safety-related regulations may also impact siting decisions and affect the operation of existing facilities. The Secretary is authorized to order corrective action if operating an LNG facility could be hazardous to life, property, or the environment (49 U.S.C. §§ 60112, 60117). DOT's implementing regulations for the Pipeline Safety Act, as amended, are in 49 C.F.R.§§ 190- 199. Safety standards, including those on siting, for LNG facilities are in 49 C.F.R. § 193 and are overseen by the Department's Office of Pipeline Safety (OPS) within the Pipeline and Hazardous Materials Safety Administration (PHMSA).

The siting provisions in 49 C.F.R. § 193 incorporate by reference standard 59A from the National Fire Protection Association (NFPA).[50] NFPA 59A requires thermal exclusion zones and flammable vapor-gas dispersion zones around LNG terminals (§§ 193.2057, 193.2059). The DOT regulations also adopt many of NFPA's design and construction guidelines including requirements for LNG facilities to withstand fire, wind, hydraulic forces, and erosion from LNG spills (§§ 193.2067, 193.2155, 193.2301). Other provisions address operations (§§ 193.2501-2521), maintenance (§§ 193.2601-2639), employee qualification (§§ 193.2701-2719), and security (§§ 193.2901-2917).

Federal Energy Regulatory Commission (FERC)

Under the Natural Gas Act of 1938 (NGA), FERC grants federal approval for the siting of new onshore LNG facilities.[51] Section 7 of the NGA authorizes FERC to issue certificates of "public convenience and necessity" for "the construction or extension of any facilities ... for the transportation in interstate commerce of natural gas" (15 U.S.C. § 717f). Section 7 does not expressly mention LNG facilities, however, so recent agency policy has FERC exercising LNG siting regulation under its Section 3 authority, which authorizes FERC to approve the import and export of natural gas (15 U.S.C. § 717b).[52] Specifically, FERC asserts approval authority over the place of entry and exit, siting, construction, and operation of new LNG terminals as well as modifications or extensions of existing LNG terminals.[53]

The Energy Policy Act of 2005 (P.L. 109-58) amends Section 3 of the NGA to give FERC explicit and "exclusive" authority to approve onshore LNG terminal siting applications (§ 311c). The 2005 act requires FERC to promulgate regulations for pre-filing of LNG import terminal siting applications and directs FERC to consult with designated state agencies regarding safety in considering such applications. It permits states to conduct safety inspections of LNG terminals in conformance with federal regulations, although it retains enforcement authority at the federal level. The 2005 act also requires LNG terminal operators to develop emergency response plans, including cost-sharing plans to reimburse state and local governments for safety and security expenditures (§ 311(d)). The 2005 act designates FERC as the "lead agency for the purposes of coordinating all applicable Federal authorizations" and for complying with federal environmental requirements, discussed below (§ 313a). It also establishes FERC's authority to set schedules for federal authorizations and establishes provisions for judicial review of FERC's siting decisions in the U.S. Court of Appeals, among other administrative provisions (§ 313(b)).

FERC implements its authority over onshore LNG terminals through the agency's regulations at 18 C.F.R. § 153. These regulations detail the application process and requirements under Section 3 of the NGA. The process begins with a pre-filing, which must be submitted to FERC at least six months prior to the filing of a formal application. The pre-filing procedures and review processes are set forth at 18 C.F.R. § 157.21. Once the pre-filing stage is completed, a formal application may be filed. FERC's formal application requirements include detailed site engineering and design information, evidence that a facility will safely receive or deliver LNG, and delineation of a facility's proposed location (18 C.F.R. § 153.8). Additional data are required if an LNG facility will be in an area with geological risk (18 C.F.R. § 153.8). The regulations also require LNG facility builders to notify landowners who would be affected by the proposed facility (18 C.F.R. § 157.6d). Facilities to be constructed at the Canadian or Mexican borders for import or export of natural gas also require a Presidential Permit.[54] According to FERC officials, applications under their Section 3-based regulations are also sufficient for Presidential Permit purposes (18 C.F.R. §§ 153.15-153.17).[55]

Under the National Environmental Policy Act of 1969 (P.L. 91-190), FERC must prepare an environmental impact statement during its review of an LNG terminal siting application (18 C.F.R. § 380.6). Applicants must prepare certain environmental reports to aid FERC in its preparation of the environmental impact statement (18 C.F.R. § 380.3(c)(2)(i)). These reports require analysis of, among other things, the socioeconomic impact of the LNG facility, geophysical characteristics of the site, safeguards against seismic risk, facility effects on air and noise quality, public safety issues in the event of accidents or malfunctions, and facility compliance with reliability standards and relevant safety standards (18 C.F.R. § 380.12). Once these environmental reports are received, the EPA may become involved in the approval process. The EPA often assists in the review of the environmental reports and the issuance of the environmental impact statements.[56]

In an effort to speed the review process for natural gas infrastructure projects (including LNG projects), FERC has approved rules to expand eligibility for "blanket certificates." Blanket certificates are granted by FERC to companies that have previously been granted certificates for construction for public convenience and necessity under Section 7 of the NGA. A company that possesses a blanket certificate may improve or upgrade existing facilities or construct certain new facilities without further case-by-case authorization from FERC. Regulations governing acceptable actions under blanket certificate authority can be found at 18 C.F.R. §§ 157.201- 157.218.

FERC also has created a Liquefied Natural Gas Compliance Branch to monitor the safety of operational LNG facilities on an ongoing basis.[57] This branch is responsible for the continued safety inspections and oversight of operating LNG facilities, and it reviews final facility design and engineering compliance with FERC orders. The staff comprises LNG engineers, civil and mechanical engineers, and other experts. The branch coordinates FERC's LNG Engineering Branch, the U.S. Coast Guard (USCG), and DOT to address safety and security at LNG facilities.[58]

FERC-DOT Jurisdictional Issues

Jurisdiction between the two federal agencies with LNG oversight responsibilities historically has been a point of contention.[59] In practice, FERC requires compliance with DOT's siting and safety regulations as a starting point, but can regulate more strictly if it chooses. This working arrangement is not explicitly established under the relevant federal law. Neither do the statutes and regulations clearly define the roles of the agencies vis-a-vis one another. The Pipeline Safety Act, for example, states:

> In a proceeding under section 3 or 7 of the Natural Gas Act (15 U.S.C. § 717b or 717f), each applicant ... shall certify that it will design, install, inspect, test, construct, operate, replace, and maintain a gas pipeline facility under ... section 60108 of this title. The certification is binding on the Secretary of Energy and the Commission... (49 U.S.C. § 60104(d)(2)).[60]

Despite this provision, which might appear to give DOT full control of gas safety regulation (including LNG siting authority), the authors of the House committee report for the revised Pipeline Safety Act indicated their intention to preserve FERC jurisdiction over LNG.[61]

Accordingly, FERC has held that the Pipeline Safety Act does not remove its jurisdiction under the NGA to regulate LNG safety.[62] In 1985, FERC and DOT executed a Memorandum of Understanding expressly acknowledging "DOT's exclusive authority to promulgate Federal safety standards for LNG facilities" but recognizing FERC's ability to issue more stringent safety requirements for LNG facilities when warranted. This agreement appears to have resolved any jurisdictional conflict between the agencies at that time.[63] In February 2004, FERC streamlined the LNG siting approval process through an agreement with the USCG and DOT to coordinate review of LNG terminal safety and security. The agreement "stipulates that the agencies identify issues early and quickly resolve them."[64]

U.S. Coast Guard

The USCG has authority to review, approve, and verify plans for marine traffic around proposed onshore LNG marine terminals as part of the overall siting approval process led by FERC. The USCG is responsible for issuing a Letter of Recommendation regarding the suitability of waterways for LNG vessels serving proposed terminals. The agency is also responsible for ensuring that full consideration is given in siting application reviews to the safety and security of the port, the LNG terminal, and the vessels transporting LNG. The USCG acts as a cooperating agency in the evaluation of LNG terminal siting applications. The Coast Guard provides guidance to applicants seeking permits for onshore LNG terminals in "Guidance on Assessing the Suitability of a Waterway for Liquefied Natural Gas Marine Traffic" (NVIC 05-05) issued on June 14, 2005.[65] Provisions in the Coast Guard Authorization Act of 2010 (H.R. 3619) would require additional waterway suitability notification requirements in LNG siting reviews by FERC (Sec. 1117).

National Fire Protection Association (NFPA)

As noted above, LNG terminal safety regulations incorporate standards set by the NFPA. The NFPA is an international nonprofit organization which advocates fire prevention and serves as an authority on public safety practices. According to NFPA, its 300 safety codes and standards "influence every building, process, service, design, and installation in the United States."[66] The NFPA LNG Standards Committee includes volunteer experts with diverse representation from industry and government, including FERC, DOT, USCG, and state agencies. The NFPA standards for LNG safety were initially adopted in 1967, with 10 subsequent revisions, most recently in 2009.[67] According to the Society of International Gas Tanker and Terminal Operators (SIGTTO), although the NFPA standards originated in the United States, they were the first internationally recognized LNG standards and are widely used throughout the world today.[68]

State Regulatory Roles

While the federal government is primarily responsible for LNG terminal safety and siting regulation, state and local laws, such as environmental, health and safety codes, can affect

LNG facilities as well. Under the Pipeline Safety Act, a state also may regulate *intra*state pipeline facilities if the state submits a certification under section 60105(a) or makes an agreement with the DOT under section 60106. Under these provisions, a state "may adopt additional or more stringent safety standards" for LNG facilities so long as they are compatible with DOT regulations (49 U.S.C. 60104(c)). Of course, if a particular LNG facility would otherwise not fall under FERC and DOT jurisdiction, states may regulate without going through the certification or agreement process. Regulation of *inter*state facilities remains the primary responsibility of federal agencies. The Office of Pipeline Safety may, however, delegate authority to *intra*state pipeline safety offices, allowing state offices to act as "agents" administering *inter*state pipeline safety programs (excluding enforcement) for those sections of *inter*state facilities within their boundaries.[69] All 50 states, the District of Columbia, and Puerto Rico are participants in the natural gas pipeline safety program.

State regulation of LNG facility safety and siting runs the gamut from piecemeal to comprehensive. For example, Arizona sets out specific requirements for LNG storage facilities, including "peak shaving" plants used by regional gas utilities, consistent with DOT regulations for construction maintenance and safety standards (Ariz. Admin. Code R14-5-202, R14-5-203, 126-01-001). Colorado and Georgia have comprehensive administrative systems for enforcing the federal standards (see 4 Co. Admin. Code 723-11; Ga. Admin. Code 515-9-3-03).

Apart from state regulation aimed specifically at LNG facilities, generally applicable state and local laws, such as zoning laws and permit requirements for water, electricity, construction, and waste disposal, also may impact the planning and development of LNG facilities. This is discussed in more detail later in this report.

Federal-State Jurisdictional Conflicts

Federal and state government agencies have had jurisdictional disagreements specifically related to the siting of new LNG terminals. In February 2004, for example, the California Public Utilities Commission (CPUC) disputed FERC's jurisdiction over the siting of a proposed LNG terminal at Long Beach because, in the CPUC's opinion, the terminal would not be involved in interstate sales or transportation and therefore would not come under the Natural Gas Act.[70] In March 2004, FERC rejected the CPUC's arguments and asserted exclusive regulatory authority for all LNG import terminal siting and construction.[71] In April 2004, the CPUC voted to assert jurisdiction over the Long Beach terminal and filed a request for FERC to reconsider its March ruling.[72] In June 2004, FERC reasserted its March ruling, prompting a federal court appeal by California regulators. The Energy Policy Act of 2005 effectively codified FERC's jurisdictional rulings, however, leading the CPUC to drop its lawsuit challenging FERC's LNG siting authority in September 2005. Notwithstanding the CPUC case, other state challenges to FERC jurisdiction remain a possibility.

KEY POLICY ISSUES

Proposals for new LNG terminal facilities have generated considerable public concern in many communities. Some community groups and government officials fear that LNG terminals may expose nearby residents to unacceptable hazards, and that these hazards may not be appropriately considered in the federal siting approval process. Ongoing public concern about LNG terminal safety has focused congressional attention on the exclusivity of FERC's LNG siting authority, proposals for a regional LNG siting process, the lack of "remote" siting requirements in FERC regulations, state permitting requirements under the Clean Water Act (CWA) and the Coastal Zone Management Act (CZMA), terrorism attractiveness of LNG, the adequacy of Coast Guard security resources, and other issues.

"Exclusive" Federal Siting Authority

As stated earlier in this report, the Energy Policy Act of 2005 (P.L. 109-58) gives FERC the "exclusive" authority to approve onshore LNG terminal siting applications (§ 311(c)). Supporters of this provision argue that it is necessary to prevent federal-state jurisdictional disputes over LNG siting authority, and that it reduces the possibility that state agencies might prevent or unduly delay the development of LNG infrastructure considered essential to the nation's energy supply. They further argue that states retain considerable influence over LNG siting approval through their federally delegated permitting authorities under the Coastal Zone Management Act of 1972 (16 U.S.C. 1451 et seq.), the Clean Air Act (42 U.S.C. 7401 et seq.), and the Federal Water Pollution Control Act (33 U.S.C. 251 et seq.). They maintain that states have a role in siting reviews under provisions in P.L. 109-58 requiring FERC to consult with governor-designated state agencies regarding state and local safety considerations prior to issuing LNG terminal permits (§ 311(d)).

A number of lawmakers at the federal and state levels have suggested that Congress should consider amending or repealing FERC's exclusive authority under P.L. 109-58. Critics of this authority argue that it vests too much power in the federal government at the expense of state agencies, which may have a better understanding of local siting issues and may bear most of the risks or burdens associated with a new LNG facility. They do not believe that FERC adequately seeks state input in its LNG siting reviews, nor adequately addresses state concerns in its siting decisions.[73] Critics question why governors lack the authority to veto onshore LNG terminal proposals as they can offshore terminal proposals under the Deepwater Port Act (33 U.S.C. § 1503(c)(8)). Some in Congress have proposed granting governors similar veto authority over onshore LNG terminal proposals, or other legislation to increase state authority in terminal siting reviews. Several legislative proposals in the 110th Congress would have required state concurrence of federal siting approval decisions for onshore LNG terminals or would have repealed provisions in the Energy Policy Act of 2005 granting FERC exclusive authority to approve LNG terminal siting applications.[74]

Regional Siting Approach

In areas such as the Northeast, where a number of onshore LNG terminal proposals have been particularly controversial, some policy makers have sought to establish a regional approach for identifying suitable sites for such terminals. They argue that FERC's consideration of LNG terminals on a proposal-by-proposal basis does not adequately take into account the regional needs for LNG, public safety concerns, and environmental impacts.[75] They also argue that the proposal-by-proposal approach does not adequately account for the relative merits of multiple LNG and natural gas pipeline facilities proposed in the same region.[76] They assert a regional LNG siting process would be more efficient than FERC's current process because it would focus attention on sites and projects with the highest chances of success rather than having numerous communities and state and local agencies react to individual plans, many of which are unlikely to be approved.[77] One legislative proposal in the 110th Congress would have established a national commission for the placement of natural gas infrastructure, such as LNG terminals, taking regional energy and environmental considerations into account.[78]

FERC officials reportedly have stated in the past that while they are not opposed to regional siting in principle, the commission cannot adopt such a regional approach because it has no land-use authority or responsibility and must let the energy market determine which terminals ultimately are constructed.[79] FERC officials also have reportedly expressed skepticism about the effectiveness of regional siting processes, for example, in finding storage locations for low-level radioactive waste.[80] More recently, however, the acting chairman of FERC reportedly called for more assessment of alternatives to LNG terminals in LNG siting decisions "including full examination of regional gas infrastructure."[81] Whether this statement indicates a future shift in FERC's approach towards LNG siting reviews remains to be determined. As oversight of federal LNG siting activities continues in the 111th Congress, legislators may be asked to consider whether incorporating regional approaches in the LNG siting process could alleviate state concerns about FERC's current process while supporting the nation's needs for new LNG infrastructure.

"Remote" Siting of LNG Terminals

The LNG safety provisions in the federal pipeline safety law require the Secretary of Transportation to "consider the ... need to encourage remote siting" of new LNG facilities (49 U.S.C. § 60103). Federal regulations contain no clear definition of what constitutes "remote" siting, relying instead on safety exclusion zones to satisfy the remoteness requirements under the Pipeline Safety Act. This regulatory alternative was criticized by the General Accounting Office (GAO) in 1979 testimony to Congress supporting remote siting in the Pipeline Safety Act:

> We believe remote siting is the primary factor in safety. Because of the inevitable uncertainties inherent in large-scale use of new technologies and the vulnerability of the facilities to natural phenomena and sabotage, the public can be best protected by placing these facilities away from densely populated areas.[82]

In 2003, Representative Edward Markey, an original sponsor of the Pipeline Safety Act, reportedly expressed concern that DOT regulations did not go far enough in complying with the congressional intent of the remote siting provisions.[83]

Industry and government officials maintain that exclusion zones do provide adequate public safety based on the current state of knowledge about LNG. They argue that LNG terminals are no longer a new technology and face far fewer operational uncertainties than in 1979. In particular, some experts believe that hazard models in the 1970s were too conservative. They believe that more recent models have led to a better understanding of the physical properties of LNG and, consequently, a better basis for design decisions affecting public safety.[84] They point out that LNG terminals like those in Everett, Massachusetts (1971); Barcelona, Spain (1969); Fezzano, Italy (1969); and Pyongtaek, Korea (1986) have been operating for decades near populated areas without a serious accident affecting the public. Of the 28 existing LNG terminals in Japan, a seismically active country, most are near major cities such as Tokyo and Osaka.[85] While the Algerian terminal accident was serious, experts point out that it did not lead to the catastrophic failure of the main LNG storage tanks and did not cause injuries to the general public. Nonetheless, some policy makers reportedly have called for amendments to federal energy law prohibiting new LNG terminals in urban and densely populated areas.[86]

Other Statutes that May Influence LNG Terminal Siting

The Energy Policy Act of 2005 (§ 311(c)) explicitly preserves states' authorities in LNG siting decisions under the Federal Water Pollution Control Act, the Coastal Zone Management Act of 1972, and other federal laws. Under the Federal Water Pollution Control Act, often referred to as the Clean Water Act (CWA), states have the authority to develop and enforce their own water quality standards.[87] Any federal permit applicant for a project that may discharge pollutants into navigable waters must provide the permitting agency with a certification from the state in which the discharge originates or will originate that the discharge is in compliance with the applicable provisions of the CWA, including the state's water quality standards.[88] States potentially could use their certification authority under the Clean Water Act to influence the siting of an LNG project by attaching conditions to the required water quality certificate or by denying certification. This certification authority has become an important tool used by states to protect the integrity of their waters. It is worth noting that the Energy Policy Act of 2005 created one potential avenue of relief for potential developers by providing for expedited review in a federal court of any order or action, or alleged inaction, by a federal or state agency acting under the authority of federal law.[89] Previously, parties seeking to challenge a state's decision regarding a water quality certificate had to do so in state court.[90]

States have been delegated authority under the Coastal Zone Management Act (CZMA, 16 U.S.C. § 1451 et seq.) which also could influence permitting of LNG terminals. Under the CZMA, applicants for federal permits to conduct activity affecting the coastal zone of a state must be certified by that state that the proposed activity is consistent with the state's federally approved coastal program.[91] A state wishing to forestall the licensing of an LNG terminal in its coastal waters could deny the certification required by the CZMA.[92] However, unlike the state-issued water quality certificates required for federal permitting by the Clean Water Act, the CZMA provides an alternative to applicants who are unable to obtain state certification. Under the CZMA, applicants may appeal the state's decision to the Department of Commerce, which may find that the activity is consistent with the objectives of the CZMA, or is otherwise necessary in the interest of national security, and thus override the state's denial of certification.[93] One analyst has suggested that there is a specific set of circumstances in which a state could create a regulatory stalemate pursuant to its CZMA authority by rejecting an application as incomplete (rather than rejecting it as improper or by failing to act). Under these circumstances the statute does not grant the Secretary of Commerce authority to review the decision. Battles between state regulatory agencies and applicants for LNG terminals have played out in this manner on at least two occasions.[94]

The discussion above suggests that authorities under the CWA and CZMA, at a minimum, give states the opportunity to have their concerns addressed when applicants seek federal approval for new LNG terminals. One legal commentator has stated that

> ultimately, while the EPAct of 2005 might have streamlined the federal [LNG siting] review process in some respects and changed the rules under which the review takes place, it has not dramatically changed the balance of power between the federal government and states.[95]

The courts addressed the potential tension between the CZMA and exclusive federal authority over LNG terminal siting in a dispute over a proposed LNG terminal in the Baltimore, MD, area. In AES Sparrows Point LNG, LLC v. Smith,[96] the U.S. District Court for the District of Maryland held that a recent amendment to the Baltimore County Zoning Regulations prohibiting the siting of an LNG facility in a particular "critical area" of the Chesapeake Bay was a part of the state's Coastal Zone Management Plan and thus not preempted by the Natural Gas Act as amended by the Energy Policy Act of 2005. The plaintiffs had claimed that the statutes explicitly gave LNG siting authority to the federal government, and thus the states could not interfere with FERC authority to rule on the plaintiff's LNG facility permit application.[97] However, on appeal the U.S. Court of Appeals for the 4th Circuit reversed the lower court's decision.[98] The appellate court ruled that the Baltimore County zoning regulation in question was not part of the state's Coastal Zone Management Plan because the regulation was never submitted to NOAA for approval.[99] The Supreme Court declined to review this decision in October of 2008.

Another statute that may have an emerging role in the LNG siting process is the Wild and Scenic Rivers Act of 1968 (WSRA).[100] The WSRA was enacted with the intention of preserving certain sections of rivers in the United States "in their free-flowing condition to protect the water quality of such rivers and to fulfill other vital national conservation purposes."[101] Under the WSRA, rivers may be designated as additions to the National Wild and Scenic Rivers system, or as potential additions to the system.[102] Designation of rivers prevents certain future development, including, potentially, projects licensed by FERC.

The WSRA explicitly prohibits FERC from licensing the construction of projects under the Federal Power Act "on or directly affecting any river which is designated ... as a component of the national wild and scenic rivers system."[103] However, projects and developments would be permitted above or below the section of the river designated under WSRA, if that development would not diminish the values present at the time of designation.[104] With regard to rivers that are designated as potential additions to the system, FERC similarly is prohibited from construction of projects along that river and other agencies are prohibited from assisting in such projects for certain periods of time after the designation to allow for the study and consideration of the river's inclusion in the system.[105]

LNG industry representatives have opined that the WSRA may be used to block LNG facility siting.[106] These representatives cited a legislative proposal in the 110th Congress to designate segments of the Taunton River in Massachusetts as "scenic and recreational" under the WSRA.[107] The industry representatives argue that this provision, if enacted, would be an obstacle towards the construction of the proposed Weaver's Cove LNG Terminal in

Massachusetts.[108] The Center for Liquefied Natural Gas described this provision as a "congressional hurdle" and said that it provided a case study of "the gauntlet of things that can be used to oppose a project."[109] As oversight of the federal LNG siting process continues, Congress may consider how federal authorities under the Energy Policy Act of 2005, the CWA, the CZMA, the WSRA, and other federal statutes fit together to achieve their various objectives.

Terror Attractiveness

Potential terrorist attacks on LNG terminals or tankers in the United States have been a key concern of policy makers because such attacks could cause catastrophic fires in ports and nearby populated areas. A 2007 report by the Government Accountability Office states that, "the ship-based supply chain for energy commodities," specifically including LNG, "remains threatened and vulnerable, and appropriate security throughout the chain is essential to ensure safe and efficient delivery."[110] Accordingly, the Coast Guard's FY2006 budget requested funding for "additional boat crews and screening personnel at key LNG hubs."[111] To date, no LNG tanker or land-based LNG facility in the world has been attacked by terrorists. However, similar natural gas and oil assets have been terror targets internationally. The Department of Homeland Security (DHS) included LNG tankers among a list of potential terrorist targets in a security alert late in 2003.[112] The DHS also reported that "in early 2001 there was some suspicion of possible associations between stowaways on Algerian flagged LNG tankers arriving in Boston and persons connected with the so-called 'Millennium Plot'" to bomb targets in the United States. Although these suspicions could not be proved, DHS stated that "the risks associated with LNG shipments are real, and they can never be entirely eliminated."[113] The 2004 report by Sandia National Laboratories concluded that potential terrorist attacks on LNG tankers could be considered "credible and possible."[114] Former Bush Administration counterterrorism advisor Richard Clarke has asserted that terrorists have both the desire and capability to attack LNG shipping with the intention of harming the general population.[115]

Although they acknowledge the security information put forth by federal agencies, some experts believe that concern about threats to LNG infrastructure is overstated.[116] In 2003, the head of one university research consortium reportedly remarked, "from all the information we have ... we don't see LNG as likely or credible terrorist targets."[117] Industry representatives argue that deliberately causing an LNG catastrophe to injure people might be possible in theory, but would be extremely difficult to accomplish. Likewise, FERC and other experts believe that LNG facilities are relatively secure compared with other hazardous chemical infrastructures that receive less public attention. In a December 2004 report, FERC stated that

> for a new LNG terminal proposal ... the perceived threat of a terrorist attack may be considered as highly probable to the local population. However, at the national level, potential terrorist targets are plentiful.... Many of these pose a similar or greater hazard to that of LNG.[118]

FERC also remarked, however, that "unlike accidental causes, historical experience provides little guidance in estimating the probability of a terrorist attack on an LNG vessel or onshore storage facility."[119] Former Director of Central Intelligence James Woolsey has stated his belief that a terrorist attack on an LNG tanker in U.S. waters would be unlikely because its potential impacts would not be great enough compared with other potential targets.[120] LNG terminal operators that have conducted proprietary assessments of potential terrorist attacks against LNG tankers have expressed similar views.[121] In its September 2006 evaluation of a proposed LNG terminal in Long Island Sound, the USCG stated that "there are currently no specific, credible threats against" the proposed LNG facility or tankers serving the facility.[122] The evaluation also noted, however, that the threat environment is dynamic and that some threats may be unknown.[123] Echoing this perspective, a 2008 report by the Institute for the Analysis of Global Security states

> Proponents are correct in that both safety and security measures currently in place make LNG terminals and ships extremely hard targets for terrorists. However, it would be imprudent to believe that terrorists are either incapable or unwilling to attack such targets. It would be equally imprudent to assume that these targets are impenetrable. If anything, in today's environment, insiders will always remain a potential threat.[124]

Because the probability of a terrorist attack on LNG infrastructure cannot be known, policy makers and community leaders must, to some extent, rely on their own judgment to decide whether LNG security is adequately addressed in FERC siting application reviews. As oversight of the federal role in LNG terminal siting continues, Congress may explore policies to reduce this uncertainty by improving the gathering and sharing of terrorism intelligence related to LNG.

Public Costs of LNG Marine Security

The potential increase in security costs from growing U.S. LNG imports, and the potential diversion of Coast Guard and safety agency resources from other activities have been a persistent concern to policy makers.[125] According to Coast Guard officials, the service's LNG security expenditures are not all incremental, since they are part of the Coast Guard's general mission to protect the nation's waters and coasts. Nonetheless, Coast Guard staff have acknowledged that resources dedicated to securing maritime LNG might be otherwise deployed for boating safety, search and rescue, drug interdiction, or other security missions.

In a December 2007 report, the GAO recommended that the Coast Guard develop a national resource allocation plan to address growing LNG security requirements.[126] In subsequent testimony before Congress, Coast Guard Commandant Admiral Thad Allen expressed concern about the costs to the Coast Guard of securing dangerous cargoes such as LNG and called for a "national dialogue" on the issue.[127] During questioning, Admiral Allen acknowledged that the Coast Guard did not currently possess sufficient resources to secure future LNG deliveries to a proposed LNG terminal in Long Island Sound which has subsequently been authorized by FERC.[128]

State and local agencies also seek more funding to offset the costs of LNG security. Addressing these concerns, the Energy Policy Act of 2005 requires private and public sector cost-sharing for LNG tanker security (§ 311d). In compliance with the act and prior FERC policy, FERC officials require new LNG terminal operators to pay the costs of any additional security or safety needed for their facilities.[129] FERC has also recommended that LNG terminal operators provide private security staff to supplement Coast Guard and local government security forces.[130] A legislative proposal in the 110th Congress would have prohibited LNG facility security plans based upon the provision of security by a state or local government lacking an LNG security arrangement with the facility operator, and would have required the Coast Guard to certify that it has adequate security resources in the sector where a terminal would be located before facility security plans for a new LNG terminal could be approved.[131] In the 111th Congress, the public provision of LNG security continues to be an issue. The Maritime Hazardous Cargo Security Act (S. 1385), introduced by Senator Lautenberg and three co-sponsors on June 25, 2009, would require a national study to identify measures to improve the security of maritime transportation of liquefied natural gas, among other provisions (Sec. 6).

Other Issues

Conducting More Safety Research

Analysts have suggested for several years that Congress could call for additional LNG safety research to help reduce uncertainties about specific LNG terminal or shipping hazards.[132] Among the LNG terminal hazard reports issued by federal agencies, LNG developers, and community groups, there appears to be widespread agreement that additional "objective" LNG safety research would be beneficial. The ABSG report states, for example, that "additional research will need to be performed to develop more refined models, and additional large-scale spill tests would be useful for providing better data for validation of models."[133] The 2004 Sandia study similarly concluded that "obtaining experimental data for large LNG spills over water would provide needed validation and help reduce modeling uncertainty."[134] Physical testing (as opposed to computer simulations) of impacts, explosions, and thermal stresses on LNG tanker hulls could also fill important gaps in engineering knowledge about the potential effects of terrorist attacks.

In 2008, Congress appropriated $8 million to fund large-scale LNG fire experiments by the Department of Energy addressing some of the hazard modeling uncertainties identified in the 2007 GAO report.[135] In that report, the GAO stated that DOE's proposed research plan at that time would "address only 3 of the top 10 issues—and not the second-highest ranked issue—that our panel of experts identified as potentially affecting public safety."[136] In response to the GAO's concerns and those of congressional staff, the DOE and Sandia modified their test program to better align with the priorities put forward by the GAO.[137] The DOE's study could, nonetheless, still be subject to the same types of technical limitations and criticisms facing existing analysis, so while it may reduce key uncertainties, it may not eliminate them altogether. As of December 2009, Sandia had completed two large LNG pool fire experiments, although the test results are not yet available.

Developer Employee Disclosure

Some policy makers have been concerned that LNG terminal developers may engage in nonpublic community lobbying or other similar activities promoting individual LNG terminals. Concern arises that these activities may limit public information and awareness about proposed terminals and, therefore, may impede the federal LNG siting review process. Accordingly, legislation proposed in the 110th Congress would have required an applicant for siting approval for an LNG terminal to identify each of its employees and agents engaged in activities to persuade communities of the benefits of the terminal.[138] Supporters of such a policy view it as a means of ensuring public transparency in LNG terminal siting. Disclosure requirements of this type might trigger some First Amendment concerns, however. The Supreme Court has recognized that such government disclosure requirements may have a deterrent effect on the exercise of First Amendment rights.[139] In balancing First Amendment interests, for example, against the government's interest preserving the integrity of the legislative process, the Court has generally upheld the constitutionality of disclosure requirements related to "direct" lobbying of members of Congress.[140] It is unclear how the Court would rule on a disclosure law such as S. 323 related to "indirect" lobbying efforts targeting constituents or otherwise taking place at the local level.[141]

Reducing LNG Demand

Some policy makers argue that Congress should try to reduce the need for new LNG terminals by acting to curb growth in domestic LNG demand, or growth in natural gas demand overall. For example, Congress could change public and industrial incentives for conservation and efficiency, switching to other fuels, or developing renewable energy supplies. Conservation and renewable energy provisions in the American Recovery and Reinvestment Act of 2009 (P.L. 111-5), which was signed by President Obama on February 17, 2009, exemplify such policies. Switching to nuclear power or biomass, however, poses its own hazards to communities and the environment, and so may not be preferable to additional LNG infrastructure. Conservation and renewable energy sources are less hazardous, although they face significant technological and cost barriers to public adoption on the scale that would be required.

Another potential way to curb U.S. LNG demand is to encourage greater North American production of natural gas. Provisions in the Energy Policy Act of 2005 promote this objective, as do proposals to encourage construction of an Alaska gas pipeline and to expand natural gas production on the outer continental shelf. An Alaska gas pipeline would take years to build, however, and might not on its own be able to meet anticipated long-term growth in U.S. gas demand.[142] Increased production from natural gas wells in the lower 48 states since 2005, as well as the recent U.S. economic recession, have reduced possible near-term pressure on natural gas supplies. It is unclear, however, if new domestic gas supplies may offer a sufficient long-term natural gas supply to meet rising gas demand in the future.

CONCLUSION

Proposals for new U.S. LNG import terminals pose safety challenges. LNG is inherently hazardous and its infrastructure is potentially attractive to terrorists. The 2004 LNG terminal fire in Algeria demonstrates that, despite technological improvements since the 1940s, LNG facilities can still experience serious accidents. Many lawmakers and the general public are concerned about these hazards.

The U.S. LNG industry is subject to more extensive siting and safety regulation than many other similarly hazardous facilities. Federal, state, and local governments have also put in place security measures intended to safeguard LNG against newly perceived terrorist threats. Some community groups and other stakeholders fear that federal siting requirements for LNG facilities are still not stringent enough, but the responsible federal agencies disagree.

The safety issues associated with LNG terminal siting are both important and familiar. Every major energy source poses some hazard to public safety. Similar public concerns have been raised around siting of other types of energy facilities such as nuclear power plants, oil import terminals, pipelines, and electric transmission lines. In evaluating new LNG terminal proposals, therefore, policy makers face a full range of facilities and safety hazards associated with U.S. energy supplies, not only LNG needs and hazards on their own.

Although LNG terminal regulations are extensive, and the global industry has decades of experience operating LNG facilities, many stakeholders question LNG terminal safety. Some of these questions might be resolved through additional research on key LNG topics. LNG siting decisions are already underway, however, so any additional research efforts intended to affect the siting process would probably have to be completed quickly. Revising LNG safety requirements after completion of a facility could be disruptive of energy supplies. Some cite the Shoreham nuclear power plant in the 1980s, which was closed after construction due to new public safety requirements, as an example of the need to resolve safety concerns before capital is invested.

Both industry and government analysts project continued growth in the demand for natural gas. Greater LNG imports represent one way to address this growth in demand, along with increased North American gas production, conservation, fuel-switching, and the development of renewable energy sources. One way or another the fundamental gas supply and demand balance must be maintained. If policy makers encourage LNG imports, then the need to foster the other energy options may be diminished—and vice versa. Thus decisions about LNG infrastructure could have consequences for a broader array of natural gas supply policies.

APPENDIX. OFFSHORE LNG TERMINAL REGULATION

Under the Deepwater Port Act of 1974 (P.L. 93-627) the Secretary of Transportation is directed to "authorize and regulate the location, ownership, construction, and operation of deepwater ports" (33 U.S.C. §§ 1501(a), 1503). The Secretary has delegated this authority to the Maritime Administration (MARAD) within the Department of Transportation, and to the Coast Guard (USCG), within the Department of Homeland Security.[143] Originally, P.L. 93-627 applied only to offshore oil ports and terminals and not LNG facilities. However, the Maritime Transportation Security Act of 2002 (P.L. 107-295) amended P.L. 93-627 to include natural gas facilities, including LNG terminals, developed offshore. As amended, "deepwater ports" are:

> any fixed or floating manmade structure other than a vessel ... located beyond State seaward boundaries ... intended for use as a port or terminal for the transportation, storage, or further handling of oil *or natural gas* for transportation to any State... (33 U.S.C. § 1502(9a))[144]

The Deepwater Port Act sets out a detailed process for offshore facility siting applications. The act also authorizes regulations addressing potential threats to the environment or human welfare posed by development of offshore LNG facilities (33 U.S.C. §§ 1504, 1508; 33 C.F.R. § 148). The act also requires regulations for the designation of safety zones around deepwater ports (33 U.S.C. § 1509(d)). Among the amendments to the act is a provision exempting LNG terminals from the limitation on the number of "deepwater ports" that can be located in a designated "application area," a provision applicable to oil terminals (33 U.S.C. §§ 1504(d)(4), (i)(4)). Additionally, a preexisting provision of the act allows the governor of a state adjacent to a proposed offshore LNG facility to have that facility license conform to state environmental protection, land and water use, or coastal zone management programs (33 U.S.C. § 1508(b)).

The USCG's regulations regarding LNG facilities are codified throughout 33 C.F.R., with major provisions in part 127. These regulations detail the requirements for siting applications, which include information about the proposed location, design, construction, and operation (33 C.F.R. § 148.109). NEPA analysis is often instrumental in siting and safety-related decisions at specific proposed facilities and is facilitated by the Minerals Management Service, the agency responsible for offshore minerals extraction and the Outer Continental Shelf leasing program.[145] Unlike requirements for onshore facilities, the Coast Guard does not appear to require generally applicable exclusion zones for offshore facilities, but relies instead on case-by-case designation of safety zones.[146] Additional USCG regulations include agency oversight of emergency procedures, security, fire protection, and design and construction standards (33 C.F.R. §§ 127.109, 127.701-127.711, 127.601-127.617, 127.1101-127.1113, 149.205).

End Notes

[1] 49 C.F.R. § 172.101. *List of Hazardous Materials.* Office of Hazardous Materials Safety, U.S. Department of Transportation.

[2] LNG proposals in the 110th Congress are found in: H.R. 1564, H.R. 2024, H.R. 2830, H.R. 6720, S. 323, S. 1174, S. 1579, S. 2822, S. 3441.

[3] Energy Information Administration (EIA). *Natural Gas Monthly.* November 2009. p. 9. Data are published only for the first nine months of the year.

[4] Energy Information Administration (EIA). *U.S. Natural Gas Imports by Country.* Online database, November 30, 2009. http://tonto.eia.doe.gov/dnav/ ng/ng_ move_impc_s1_m.htm.

[5] Energy Information Administration (EIA). *Natural Gas Annual 2005.* Tables 1 and 9. November 16, 2006.

[6] Energy Information Administration (EIA). "World LNG Imports by Origin, 2002." Washington, DC. October 2003.

[7] Greenspan, A., Chairman, U.S. Federal Reserve Board. "Natural Gas Supply and Demand Issues." Testimony before the House Energy and Commerce Committee. June 10, 2003.

[8] *Inside F.E.R.C.* "Kelly-LNG Poised for 'Major Contribution' to Energy Supply, to Meet Industrial Demand." April 10, 2006.

[9] See, for example: John D. Podesta and Timothy E. Wirth. *Natural Gas: A Bridge Fuel for the 21st Century.* Center for American Progress. August 10, 2009. http://www.americanprogress.org/issues/ 2009/08/pdf/ natural gasmemo.pdf.

[10] Rita Tubb. "LNG Imports Projected to Rise." *Pipeline & Gas Journal.* May 1, 2009.

[11] Federal Energy Regulatory Commission, "Proposed North American LNG Import Terminals," October 27, 2009. http://www.ferc.gov/industries/lng/indus-act/terminals/lng-proposed.pdf.

[12] Bureau of Mines (BOM). *Report on the Investigation of the Fire at the Liquefaction, Storage, and Regasification Plant of the East Ohio Gas Co., Cleveland, Ohio, October 20, 1944.* February 1946.

[13] Junnola, Jill, et al. "Fatal Explosion Rocks Algeria's Skikda LNG Complex." *Oil Daily*. January 21, 2004. p. 6.
[14] Methane, the main component of LNG, burns in gas-to-air ratios between 5% and 15%.
[15] Havens, J. "Ready to Blow?" *Bulletin of the Atomic Scientists*. July/August 2003. p. 17.
[16] Havens. 2003. p. 17.
[17] West, H.H. and Mannan, M.S. "LNG Safety Practices and Regulations." Prepared for the American Institute of Chemical Engineering Conference on Natural Gas Utilization and LNG Transportation. Houston, TX. April 2001. p. 2.
[18] Quillen, D. ChevronTexaco Corp. "LNG Safety Myths and Legends." Presentation to the Natural Gas Technology Conference. Houston, TX. May 14-15, 2002. p. 18.
[19] Siu, Nathan, et al. *Qualitative Risk Assessment for an LNG Refueling Station and Review of Relevant Safety Issues*. Idaho National Engineering Laboratory. INEEL/EXT-97-00827 rev2. Idaho Falls, ID. February 1998. p. 71.
[20] Siu. 1998. p. 62.
[21] Siu. 1998. p. 63.
[22] Quillen. 2002. p. 28.
[23] Skolnik, Sam. "Local Sites Potential Targets for Cyberterror." *Seattle Post-Intelligencer*. Seattle, WA. September 2, 2002.
[24] Center for LNG. "LNG Carrier Safety: A Long Record of Safe Operation." Internet page., December 14, 2009. http://www.lngfacts.org/About-LNG/Carrier-Safety.asp; Groupe International des Importateurs de Gaz Naturel Liquéfié, "LNG Ships," Paris, France, 2009, p. 1, http://www.giignl.org/fileadmin/user_upload/pdf/LNG_Safety/3%20-%20LNG%20Ships%208.28.09%20Final%20HQ.pdf.
[25] Foss, 2007; CH-IV International. *Safety History of International LNG Operations*. TD-02109. Millersville, MD. December. pp. 13-18.
[26] Society of International Gas Tanker & Terminal Operators Ltd. (SIGTTO). "Safe Havens for Disabled Gas Carriers." Third Edition. London. February 2003. pp. 1-2.
[27] Petroplus International, N.V. "Energy for Wales: LNG Frequently Asked Questions." Internet home page. Amsterdam, Netherlands. August 4, 2003.
[28] Junnola, J., et al. January 21, 2004. p. 6.
[29] Hunter, C. "Algerian LNG Plant Explosion Sets Back Industry Development." *World Markets Analysis*. January 21, 2004. p. 1.
[30] Antosh, N. "Vast Site Devastated." *Houston Chronicle*. January 21, 2004. p. B1.
[31] See, for example: Jerry Havens, *Comments Submitted by Jerry Havens on Sparrows Point Final Environmental Impact Statement*, Federal Energy Regulatory Commission, Docket CP07-62-000, January 12, 2009.
[32] Gas Research Institute (GRI). "LNGFIRE: A Thermal Radiation Model for LNG Fires" Version 3. GRI-89/0176. Washington, DC. June 29, 1990; "LNG Vapor Dispersion Prediction with the DEGADIS: Dense Gas Dispersion Model." GRI-89/00242; "Evaluation of Mitigation Methods for Accidental LNG Releases. Vol. 5: Using FEM3A for LNG Accident Consequence Analyses." GRI 96/0396.5. Washington, DC.
[33] FERC. November 2003. p. 4-133.
[34] ABSG Consulting. May 13, 2004. p. iii.
[35] SNL. December 2004. p. 14.
[36] SNL. December 2004. p. 14.
[37] Raj, P.K. *Spectrum of Fires in an LNG Facility: Assessments, Models and Consideration in Evaluations*. Prepared for the U.S. Department of Transportation, Pipeline & Hazardous Materials Safety Admin. by Technology & Management Systems, Inc. Burlington, MA. December 5, 2006. p. E-4.
[38] Government Accountability Office (GAO). *Maritime Security: Public Safety Consequences of a Terrorist Attack on a Tanker Carrying Liquefied Natural Gas Need Clarification*. GAO-07-316. February 2007. p. 22.
[39] See, for example Senator Barbara A. Mikulski, testimony before the House Transportation and Infrastructure Committee, Coast Guard and Maritime Transportation Subcommittee field hearing on the Safety and Security of Liquefied Natural Gas and the Impact on Port Operations. Baltimore, MD. April 23, 2007.
[40] One attempt at such a study is: Clarke, R.A., et al. *LNG Facilities in Urban Areas*. Good Harbor Consulting, LLC. Prepared for the Rhode Island Office of Attorney General. GHC-RI-0505A. May 2005.
[41] Based on facilities submitting Risk Management Plans required under Section 112 of the Clean Air Act (42 U.S.C. § 7412) and classified in the December 1, 2003, update of the EPA National Database using EPA's software RMP*Review (v2.1). EPA states that an entire population is highly unlikely to be affected by any single chemical release, even in the worst case. In an actual release, effects on a population would depend on

wind direction and many other factors. In addition, these worst-case scenarios do not account for emergency response measures facility operators or others might take to mitigate harm.

[42] Office of Hazardous Materials Safety, Department of Transportation. *Hazardous Materials Shipments.* Washington, DC. October 1998. Table 2. p. 2.

[43] Exxon Shipping Co. et al. v. Baker et al. 554 U.S. (2008) (Supreme Court of the United States of America). June 25, 2008. p. 4.

[44] *Energy Daily.* "Shell, Olympic Socked for Pipeline Accident." January 22, 2003.

[45] Heilprin, John. "Ashcroft Promises Increased Enforcement of Environmental Laws for Homeland Security." Associated Press. *Washington Dateline.* Washington, DC. March 11, 2003.

[46] "Massachusetts LNG Company Faces RSPA Fines for Security Violations." Bulk Transporter. June 28, 2002; Pipeline and Hazardous Materials Safety Admin. "Summary of Enforcement Actions: Distrigas of Massachusetts Corp." Web page. May 8, 2008. http://primis.phmsa.dot.gov/comm/reports/enforce/Actions_opid_3411.html#_TP_1_tab_1

[47] Business Editors. "Olympic Pipe Line, Others Pay Out Record $75 Million in Pipeline Explosion Wrongful Death Settlement." *Business Wire.* April 10, 2002.

[48] National Transportation Safety Board (NTSB). *Pipeline Accident Report* PAR-03-01. February 11, 2003.

[49] El Paso Corp. *Quarterly Report Pursuant to Section 13 or 15(d) of the Securities Exchange Act of 1934.* Form 10-Q. For the period ending June 30, 2002. Houston, TX.

[50] National Fire Protection Association (NFPA). *Standard for the Production, Storage, and Handling of Liquefied Natural Gas,* 2001 Edition. NFPA 59A. Quincy, MA. 2001.

[51] Natural Gas Act (NGA) of June 21, 1938, ch. 556, 52 Stat. 812 (codified as amended at 15 U.S.C. §§ 717 et seq.); the Department of Energy Organization Act of 1977 (P.L. 95-91) transferred to the NGA authority to approve siting, construction and operation of onshore LNG facilities to the Secretary of Energy (§ 301b). The Secretary, in turn, delegated this authority to FERC.

[52] In 1997, FERC reaffirmed its Section 3 authority despite changes to the Natural Gas Act in the Energy Policy Act of 1992 (P.L. 102-486). For details *see* 97 FERC ¶ 61,231 (2001). Also note that FERC's regulatory power regarding LNG importation under section 3 has been held to allow FERC to impose requirements equivalent to any in section 7, so long as FERC finds them necessary or appropriate to the public interest. Distrigas Corp. v. FPC, 495 F.2d 1057, 1066 (D.C. Cir. 1974).

[53] *See* 18 C.F.R. § 153; *see also* Foley, R., Federal Energy Regulatory Commission (FERC), Office of Energy Projects. "Liquefied Natural Gas Imports." Slide presentation. January 2003. p. 10.

[54] Executive Order No. 10,485 requires that FERC obtain a favorable recommendation from the Secretaries of State and Defense prior to issuing a Presidential Permit.

[55] FERC Office of Energy Projects, Personal communication, December 10, 2003.

[56] In July of 2006 EPA issued a "Liquefied Natural Gas Regulatory Roadmap" in an effort to assist LNG project applicants (both onshore and offshore) in dealing with environmental regulatory requirements. http://www.epa.gov/opei/lng/lngroadmap.pdf

[57] Press Release, Federal Energy Regulatory Commission, "Commission Establishes LNG Compliance Branch," May 2, 2006.

[58] *Id.*

[59] S.Rept. No. 96-182. 1979. p. 4.

[60] 49 U.S.C. § 61018 specifies DOT's requirements for pipeline facility inspection and maintenance.

[61] *See* H.Rept. No. 1390, 1968, *reprinted in* 1968 U.S.C.C.A.N. 3223, 3251. Note, FERC was known as the Federal Power Commission (FPC) at the time.

[62] Chatanooga Gas Co., 51 FPC 1278, 1279 (1974).

[63] See "Notice of Agreement Regarding Liquefied Natural Gas," 31 FERC ¶ 61,232 (1985).

[64] Federal Energy Regulatory Commission (FERC). Press release R-04-3. February 11, 2004.

[65] Available at http://www.uscg.mil/hq/g-m/nvic/NVIC%2005-05.doc.pdf.

[66] National Fire Protection Assoc. (NFPA). *About NFPA.* Web page. Quincy, MA. 2009.

[67] National Fire Protection Association (NFPA). *Standard for the Production, Storage, and Handling of Liquefied Natural Gas,* 2009 Edition. NFPA 59A. Quincy, MA. 2009.

[68] Society of International Gas Tanker and Terminal Operators (SIGTTO). Personal communication. London, England. December 19, 2003.

[69] 49 U.S.C. § 601. States may recover up to 50% of their costs for these programs from the federal government.

[70] FERC. "Notice of Intervention and Protest of the Public Utilities Commission of the State of California." Docket No. CP04-58-000. February 23, 2004. p. 6.

[71] Lorenzetti, M. "LNG Rules." *Oil & Gas Journal*. April 5, 2004. p. 32.

[72] *Gas Daily*. "PUC Seeks Rehearing of FERC's Order on Long Beach LNG Project." April 27, 2004. p. 7.

[73] O'Neill, L. "Senators Drum Up Measure Giving States Authority Over LNG Siting." *Natural Gas Week*. April 14, 2008.

[74] S. 1174, S. 3441, S. 2822, and H.R. 2042

[75] *Foster Natural Gas Report*, "Weaver's Cove Offshore Berth Project Must Be Reviewed From Scratch And By Taking A Regional Approach, Some Parties Insist," July 18, 2008, p. 11.

[76] See, for example: Sickinger, T. "Governor Ups Ante Against LNG Sites." *The Oregonian*. February 15, 2008. p. A1.

[77] *Bangor Daily News*. "Regional Energy." December 8, 2006. p. A10.

[78] H.R. 6720.

[79] *Foster Natural Gas Report*. "Northeast States Need LNG, Especially New England; No Evidence of Regional Planning." December 15, 2006. p. 5.

[80] Howe, P.J. "LNG Supplies, Solutions Lacking." *The Boston Globe*. September 26, 2006. p. C4.

[81] *Foster Natural Gas Report*, "Under FERC's New Acting Chairman, Renewable Energy And Efficiency May Command An Unprecedented Level Of Support," February 17, 2009, p. 12.

[82] Peach, J.D. General Accounting Office (GAO), Director, Energy and Minerals Division. Testimony to the Senate Committee on Commerce, Science and Transportation. Washington, DC. April 25, 1979. p. 10. The General Accounting Office is now known as the Government Accountability Office.

[83] Raines, B. "Congress Wanted LNG Plants at 'Remote' Sites." *Mobile Register*. Mobile, AL. November 16, 2003.

[84] Tobin, L.T. Remarks at a meeting of the City of Vallejo Seismic Safety Commission. Meeting minutes. Vallejo, CA. September 11, 2003. p. 5; see also: *Federal Register*. Vol. 62, No. 37. February 25, 1997. pp. 8402-8403.

[85] California Energy Commission (CEC). "Asia Pacific Countries Liquefied Natural Gas (LNG)." January 2008. http://www.energy.ca.gov/lng/worldwide/asia_pacific.html

[86] *Energy Washington Week*. January 10, 2007.

[87] 33 U.S.C. § 1251(a),(b).

[88] *Id.* at § 1341(a).

[89] EPAct of 2005, P.L. 109-58 at §§ 717r(d)(1)-(2).

[90] For further discussion of the water quality certification process and its impact on LNG siting, see Dweck, J., Wochner, D., and Brooks, M. "Liquefied Natural Gas (LNG) Litigation after the Energy Policy Act of 2005: State Powers in LNG Terminal Siting." *Energy Law Journal*, Vol. 27, No. 45 (2006). p. 482-85.

[91] 16 U.S.C. § 1456(c)(3)(A).

[92] The state must actively state its objection to the applicant's certification; a state's failure to act is presumed to be concurrence with project certification. *Id.*

[93] 16 U.S.C. § 1456(c)(3)(A).

[94] For further discussion, see Ewing, K.A. and E. Petersen. "Significant Environmental Challenges to the Development of LNG Terminals in the United States." *Texas Journal of Oil, Gas and Energy Law*. November 2006. pp. 21-23; Dweck et al., p. 487-90.

[95] Dweck et al. 2006, p. 475.

[96] 539 F.Supp.2d 788 (D. Md 2007).

[97] The same court had previously overturned a Baltimore County zoning ordinance that prohibited siting of LNG facilities within a certain distance of residential and commercial facilities on those same grounds. See AES Sparrows Point LNG, LLC v. Smith, 470 F.Supp.2d 586 (D.Md. 2007).

[98] AES Sparrows Point LNG, LLC v. Smith, 5237 F.3d 120 (4th Cir. 2008).

[99] Id. at 126-27.

[100] P.L. 90-542, codified at 16 U.S.C. § 1271 et seq. For an examination of the purposes, language and legislative history of this act, and an analysis of its effect on water rights, see CRS Report RL30809, *The Wild and Scenic Rivers Act (WSRA) and Federal Water Rights*, by Cynthia Brougher.

[101] 16 U.S.C. § 1271.

[102] Rivers designated as "potential additions" are those that warrant further study before full extension of the protections of the WSRA. 16 U.S.C. § 1276.

[103] 16 U.S.C. § 1278(a).

[104] Id.

[105] 16 U.S.C. § 1278(b).

[106] Clarke, D., *LNG Sector Fears Use of New Legislative Tactic to Oppose Facilities*, Energy Washington Week, July 30, 2008.

[107] Ibid.; H.R. 415.
[108] Ibid.
[109] Ibid.
[110] Government Accountability Office (GAO). "Maritime Security: Federal Efforts Needed to Address Challenges in Preventing and Responding to Terrorist Attacks on Energy Commodity Tankers," GAO-08-141, December 10, 2007, p.77.
[111] Department of Homeland Security (DHS). *Budget-in-Brief, Fiscal Year 2006.* https://www.dhs.gov/xlibrary/assets/Budget_BIB-FY2006.pdf.
[112] Office of Congressman Edward J. Markey. Personal communication with staff. January 5, 2004.
[113] Turner, Pamela J., Assistant Secretary for Legislative Affairs, Department of Homeland Security (DHS). Letter to U.S. Representative Edward Markey. April 15, 2004. p. 1.
[114] Sandia National Laboratories (SNL). *Guidance on Risk Analysis and Safety Implications of a Large Liquefied Natural Gas (LNG) Spill Over Water.* SAND2004-6258. Albuquerque, NM. December 2004. pp. 49-50.
[115] Clarke, Richard A. et al. *LNG Facilities in Urban Areas.* Good Harbor Consulting, LLC. Prepared for the Rhode Island Office of Attorney General. GHC-RI-0505A. May 2005.
[116] McLaughlin, J. "LNG Is Nowhere Near as Dangerous as People Are Making it Out to Be." *Lloyd's List.* February 8, 2005. p. 5.
[117] Behr, Peter. "Higher Gas Price Sets Stage for LNG." *Washington Post.* July 5, 2003. p. D10.
[118] Federal Energy Regulatory Commission (FERC). *Vista del Sol LNG Terminal Project, Draft Environmental Impact Statement.* FERC/EIS-0176D. December 2004. p. 4-162.
[119] FERC. FERC/EIS-0176D. December 2004. p. 4-162. Notwithstanding this assertion, in its subsequent draft review of the Long Beach LNG terminal proposal, the FERC states that "the historical probability of a successful terrorist event would be less than seven chances in a million per year...." See FERC. October 7, 2005. p. ES-14.
[120] Woolsey, James. Remarks before the National Commission on Energy LNG Forum, Washington, DC, June 21, 2006.
[121] Grant, Richard, President, Distrigas. Testimony before the Senate Committee on Energy and Natural Resources, Subcommittee on Energy hearing on "The Future of Liquefied Natural Gas: Siting and Safety." February 15, 2005.
[122] U.S. Coast Guard. *U.S. Coast Guard Captain of the Port Long Island Sound Waterways Suitability Report for the Proposed Broadwater Liquefied Natural Gas Facility.* September 21, 2006. p. 146.
[123] Ibid.
[124] Cindy Hurst. *The Terrorist Threat to Liquefied Natural Gas: Fact or Fiction?* Institute for the Analysis of Global Security. Washington, DC. p. 3.
[125] See, for example, Representative Peter Defazio, remarks before the House Homeland Security Committee hearing on Securing Liquid Natural Gas Tankers to Protect the Homeland. March 21, 2007.
[126] Government Accountability Office. *Maritime Security: Federal Efforts Needed to Address Challenges in Responding to Terrorist Attacks on Energy Commodity Tankers*. GAO-08-141. December 10, 2007. p. 79.
[127] Admiral Thad Allen, Commandant, U.S. Coast Guard. Testimony before the House Committee on Appropriations, Subcommittee on Homeland Security hearing, "Coast Guard Budget: Impact on Maritime Safety, Security, and Environmental Protection." March 5, 2008.
[128] Admiral Thad Allen, March 5, 2008; Federal Energy Regulatory Commission. "Order Granting Authority Under Section 3 of the Natural Gas Act and Issuing Certificates." Docket No. CP06-54-0000. March 20, 2008.
[129] Baldor, L.C. "Federal Agency, R.I. Officials Meet over LNG Terminal." *Associated Press.* March 17, 2005.
[130] Federal Energy Regulatory Commission (FERC). "Response to Senator Jack Reed's 2/1/05 letter regarding the proposed Weaver's Cove LNG Project in Fall River, MA & the proposed KeySpan LNG Facility Upgrade Project in Providence, RI under CP04-293 et al." March 3, 2005. p. 2.
[131] H.R. 2830.
[132] See, for example: Kytömaa, H. and Gavelli, F. "Studies of LNG Spills Over Water Point Up Need for Improvement." *The Oil and Gas Journal.* May 9, 2005. p. 61.
[133] ABSG Consulting. May 13, 2004. p. iv.
[134] SNL. December 2004. p. 18.
[135] Consolidated Appropriations Act, 2008 (P.L. 110-161), Division C—Energy and Water Development and Related Agencies Appropriations Act, 2008, Title III, Explanatory Statement, p. 570.
[136] GAO. 2007. pp. 22-23.

[137] Anay Luketa, Sandia National Laboratories. *DOE/Sandia National Laboratories Coordinated Approach for LNG Safety and Security Research*. Presentation to the Committee on Gas, NARUC Summer Committee Meetings. Portland, OR. July 22, 2008. Physical testing and model development by Sandia are scheduled to be completed in 2009.

[138] S. 323

[139] *Buckley v. Valeo*, 424 U.S. 1, 65 (1976).

[140] *United States v. Harriss*, 347 U.S. 612 (1954).

[141] For further discussion on this topic, see CRS Report RL33794, *Grassroots Lobbying: Constitutionality of Disclosure Requirements*, by Jack Maskell.

[142] For further analysis, see CRS Report R40963, *The Alaska Natural Gas Pipeline: Background, Status, and Issues for Congress*, by Paul W. Parfomak.

[143] For a recent LNG siting application, MARAD performed financial analysis and USCG evaluated environmental impacts; the agencies cooperated on all other aspects of the review. ("First Offshore Terminal in U.S. is About to Secure Federal License." *Foster Natural Gas Report*. Bethesda, MD. November 20, 2003. p. 21.

[144] The statute defines natural gas to include "liquefied natural gas." 33 U.S.C. § 1502(14).

[145] Sierra B. Weaver, Note, "Local Management of Natural Resources: Should Local Governments Be Able to Keep Oil Out?," 26 *HARV. ENVTL. L. REV.* 231, 246 (2002).

[146] See 33 C.F.R. § 165, Regulated Navigation Areas and Limited Access Areas.

Chapter 6

NATURAL GAS PASSENGER VEHICLES: AVAILABILITY, COST, AND PERFORMANCE[*]

Brent D. Yacobucci

SUMMARY

Higher gasoline prices in recent years and concerns over U.S. oil dependence have raised interest in natural gas vehicles (NGVs). Use of NGVs for personal transportation has focused on compressed natural gas (CNG) as an alternative to gasoline. Consumer interest has grown, both for new NGVs as well as for conversions of existing personal vehicles to run on CNG. This report finds that the market for natural gas passenger vehicles will likely remain limited unless the differential between natural gas and gasoline prices remains high in order to offset the higher purchase price for an NGV. Conversions of existing vehicles will also continue to be restricted unless the Clean Air Act (CAA) is amended or if the Environmental Protection Agency (EPA) makes changes to its enforcement of the CAA.

INTRODUCTION

Congressional and consumer interest in natural gas vehicles (NGVs) for personal transportation has grown in recent years, especially in response to higher gasoline prices, concerns over the environmental impact of petroleum consumption for transportation, and policy proposals such as the "Pickens Plan."[1] Although natural gas passenger vehicles have been available for years, they have been used mostly in government and private fleets; very few have been purchased and used by consumers. Larger NGVs—mainly transit buses and delivery trucks—also play a role in the transportation sector, especially due to various federal, state, and local incentives for their use. However, high up-front costs for new NGVs, as well

---

[*] This is an edited, reformatted and augmented version of a Congressional Research Services publication, dated February 3, 2010.

as concerns over vehicle performance and limited fuel infrastructure, have led to only marginal penetration of these vehicles into the personal transportation market.

CURRENT MARKET

The Energy Information Administration (EIA) estimates that there were roughly 114,000 compressed natural gas (CNG) vehicles in the United States in 2007, and roughly 3,000 liquefied natural gas (LNG) vehicles.[2] Roughly two-thirds of NGVs are light-duty (i.e., passenger) vehicles. This compares to roughly 240 million conventional (mostly gasoline) light-duty vehicles.[3] Further, of the roughly 16.1 million new light-duty vehicles sold in 2007, only about 1,100 (0.01%) were NGVs.[4] For model year (MY) 2010, only one NGV was available from an original equipment manufacturer (OEM) for purchase by consumers—the CNG-fueled Honda Civic GX[5]—although some companies convert vehicles to CNG before they are sold (usually as fleet vehicles).

LIFE-CYCLE COST ISSUES

Currently, natural gas vehicles are significantly more expensive than comparable conventional vehicles. For example, the incremental price between a conventional Honda Civic EX and a natural gas-powered Honda Civic GX is nearly $6,000,[6] although some of this difference is made up through a tax credit for the purchase of new alternative fuel vehicles. If a taxpayer qualifies, he or she may claim a credit of up to $4,000 for the purchase of a new Honda Civic GX.[7] This tax credit is set to terminate at the end of 2010. It should be noted that with higher production, this incremental cost should decrease, but the likely extent of that decrease is unclear.

Since the number of natural gas refueling stations is limited—only about 400 to 500 publicly available nationwide,[8] compared to roughly 120,000 retail gasoline stations[9]—the purchaser of a new NGV might also choose to install a home refueling system. According to *Consumer Reports* and Natural Gas Vehicles for America (NGVAmerica), a FuelMaker Phill system costs between $3,400 and $4,500 plus installation.[10] However, a taxpayer can offset $1,000 of this by claiming a tax credit for installing new alternative fuel refueling infrastructure.[11]

Offsetting the higher up-front costs are likely annual fuel savings in switching from gasoline to natural gas. Using recent average retail gasoline and residential natural gas prices, annual fuel cost savings could be roughly $650.[12] Assuming a 7% discount rate, the current payback period for the CNG vehicle and home refueling system is just over 10 years. Depending on how long a consumer keeps a new vehicle, this payback period may or may not be acceptable to that consumer.

Assuming a smaller differential between natural gas and gasoline prices, or the expiration of the existing tax incentives can significantly increase this payback period; assuming a larger difference in fuel prices, assuming a smaller discount rate, or assuming incremental natural gas vehicle prices decrease in the future, this payback period could be shorter.

OTHER POTENTIAL BENEFITS AND COSTS

In addition to the life-cycle cost difference between CNG and conventional vehicles, there are other costs and benefits associated with natural gas vehicles which may not have a defined price tag. For example, any reduction in petroleum dependence (beyond the per-gallon cost savings) is not represented in the above payback period estimate. Some consumers may place a value on displacing petroleum consumption, and thus imports.[13] Further, natural gas vehicles in general have lower pollutant and greenhouse gas emissions than comparable gasoline vehicles, although this may or may not be true for specific vehicles and pollutants.[14]

A key potential benefit raised by proponents of NGVs is that while the United States imports the majority of the petroleum it uses, most natural gas is domestically produced. Further, domestic output is higher than once thought, likely due to recent growth in unconventional natural gas sources (e.g., coal mine methane, shale gas).[15]

But there are also several potential and measurable drawbacks to natural gas vehicles, many related to vehicle performance and acceptability. For example, CNG engines tend to generate less power for the same size engine than gasoline engines. Thus NGVs tend to have slower acceleration and less power climbing hills.[16] Also, because CNG has a lower energy density than gasoline, CNG vehicles tend to have a shorter range than comparable gasoline vehicles.[17] In addition, for passenger vehicles, the larger natural gas storage tanks often occupy space that would otherwise be used for cargo—generally in the trunk of a sedan and in the bed of a pickup truck.[18] Again, these considerations may or may not play into a individual purchaser's decision, but could affect the overall marketability of the vehicles.

NGV CONVERSIONS

A key question raised by those interested in the expansion of natural gas for automobiles is whether existing vehicles can be converted to operate on natural gas. From a technical feasibility standpoint, there are few problems with converting a vehicle to operate on natural gas. Most existing engines can operate on the fuel, and most conversions involve changes to the fuel system, including a new fuel tank, new fuel lines, and modifications to the vehicle's electronic control unit.[19]

However, converting an existing vehicle is more problematic from a practical standpoint. In the United States, NGV conversions—or any other fuel conversion—can potentially run afoul of the Clean Air Act (CAA). All new vehicles (gasoline or otherwise) must pass rigorous tests to prove they will meet emissions standards over the life of the vehicle. These tests tend to be very expensive, although the marginal cost spread over a full product run—thousands to hundreds of thousands of vehicles—is minimal. After a vehicle has been certified by the Environmental Protection Agency (EPA), any changes to the exhaust, engine, or fuel systems may be considered tampering under the CAA. Section 203(a)(3)(A)[20] states that it is prohibited

> for any person to remove or render inoperative any device or element of design installed on or in a motor vehicle or motor vehicle engine in compliance with regulations under this title prior to its sale and delivery to the ultimate purchaser, or for any person

knowingly to remove or render inoperative any such device or element of design after such sale and delivery to the ultimate purchaser.

EPA generally interprets this to mean that any change to a vehicle's engine or fuel systems that leads to higher pollutant emissions constitutes "tampering" under Section 203.

In 1974, EPA issued guidance ("Memorandum 1A") to automaker and auto parts suppliers on what constituted tampering in terms of replacement parts under routine maintenance.[21] The guiding principle EPA has used in enforcing the anti-tampering provisions for alternative fuel conversions is that such changes are allowed as long as the dealer has "reasonable basis" to believe that emissions from the vehicle will not increase after the conversion. Instead of requiring all converted vehicles to undergo testing equivalent to new vehicle testing, EPA allowed vehicle converters flexibility in certifying their emissions.

However, in the 1990s, EPA received data from the National Renewable Energy Lab that many vehicles converted to run on natural gas or liquified petroleum gas (LPG) and certified under the flexibility provisions might be exceeding emissions standards.[22] Therefore, in 1997 EPA issued an addendum to Memorandum 1A tightening the testing standards for these conversions. The original decision required compliance with new testing procedures starting in 1999. Subsequent revisions extended the deadline thorough March 2002.

Currently, certifying vehicle conversions can be very expensive for small producers, since each vehicle must be independently certified. For example, a converter must test the emissions of the conversion of specific "engine families" (e.g., MY2008 4.6L V8 Ford vehicles). Each different engine/emissions system combination must be tested independently (e.g., MY2009 vehicles, or vehicles with different engines). Therefore, the production and use of universal "conversion kits" is effectively prohibited under the EPA enforcement guidance.[23] To allow a market for conversion kits, the CAA would need to be amended to allow for these conversions regardless of vehicle emissions, or EPA would need to conclude that the conversions do not increase emissions. NGVAmerica estimates that it can cost as much as $200,000 to design, manufacture, and certify a conversion for a single engine family.[24]

Some companies have completed the required testing on a limited number of vehicles, and offer conversions. NGVAmerica maintains a list of the companies that currently sell NGV conversion equipment, and the vehicles that have been certified by those companies.[25] In addition to the Civic GX produced by Honda, NGVA lists six companies that convert Ford and General Motors vehicles—mostly light-duty trucks such as pickups and vans. According to EPA requirements, vehicles must be converted by the original manufacturer of the conversion equipment, or by a retrofitter trained and qualified by the conversion manufacturer. NGVAmerica estimates that converting a passenger vehicle can cost over $10,000 (e.g., they estimated $13,500 for a Ford Crown Victoria), although specific costs would be determined by the manufacturer and/or retrofitter.[26] Conversions would be eligible for the $4,000 alternative fuel vehicle tax credit (see "Life-Cycle Cost Issues" above).

Some have questioned whether a vehicle conversion would void the original manufacturer's warranty. However, only those vehicle systems directly modified by the conversion would raise warranty concerns. In those cases, the conversion manufacturer's warranty would warranty the modified systems. For systems not affected by the conversion (e.g., suspension, climate control), the original manufacturer's warranty would still apply.[27]

LEGISLATION

Several bills have been introduced in the 111th Congress that would promote natural gas vehicles and NGV infrastructure. Most notably, the New Alternative Transportation to Give Americans Solutions Act (Nat Gas Act) of 2009 (H.R. 1835 and S. 1408) would provide a wide range of incentives. The Nat Gas Act would significantly expand tax credits for the purchase of NGVs and for the installation of natural gas refueling infrastructure, and extend those credits through 2017 (they are set to expire at the end of 2010). The bill would also provide a tax credit to automakers who produce NGVs, and would authorize grants to those automakers to develop natural gas engines. Finally, the bill would require that 50% of vehicles purchased by federal agencies be NGVs. As of February 2010, both the House and Senate versions of the bill had been referred to committee.

Several other bills would also provide additional tax incentives or government mandates for the purchase of alternative fuel vehicles, including NGVs. Other than the American Clean Energy and Security Act (ACES; H.R. 2454), the House energy and climate change bill, none of these bills has been reported out of committee. ACES would provide many incentives for the use of natural gas over other, more carbon-intensive fuels (i.e., coal and petroleum). ACES would also require a study by EPA on the potential for NGVs to reduce greenhouse gas emissions and criteria pollutants under the CAA.

CONCLUSION

Higher gasoline prices and concerns about U.S. oil dependence have raised interest in NGVs. Energy policy proposals such as the Pickens Plan have further raised interest in these vehicles. However, currently the number of new passenger vehicles capable of operating on natural gas is relatively low, and there are limited opportunities for converting existing gasoline vehicles to run on natural gas.

The market for natural gas vehicles will likely remain limited unless the differential between natural gas and gasoline prices remains high in order to offset the higher purchase price for a natural gas vehicle. Conversions of existing vehicles will also continue to be restricted unless the CAA is amended or if EPA makes changes to its enforcement of the CAA.

End Notes

[1] On July 8, 2008, T. Boone Pickens announced a plan calling for reduced petroleum imports through the expanded use of natural gas in transportation. For an analysis of this plan, see CRS General Distribution Memoradum, *The T. Boone Pickens Energy Plan: A Preliminary Analysis of Implementation Issues*, by Jeffrey Logan, William F. Hederman, and Brent D. Yacobucci.

[2] U.S. Energy Information Administration (EIA), *Alternatives to Traditional Transportation Fuels 2007*, April 2009. Tables V1 and V4.

[3] Stacy C. Davis, Susan W. Diegel, and Robert G. Boundy, *Transportation Energy Data Book: Edition 27*, 2008. Tables 4.1 and 4.2.

[4] Davis, et al., op. cit. Tables 4.5 and 4.6; EIA, op. cit. Table S1.

[5] For MY2004, there were eight CNG models (from Ford, General Motors, and Honda). This number dropped to five in MY2005, and one in MY2006. U.S. Department of Energy (DOE), *Fueleconomy.gov Website.* Accessed February, 2, 2010.

[6] 2010 Honda Civic EX MSRP: $19,455. 2010 Honda Civic GX MSRP: $25,340. $25,340 -$19,455 = $5,885.

[7] Energy Policy Act (EPAct) of 2005. P.L. 109-58, Sec. 1341.

[8] Roughly half of the 800 to 1,000 natural gas refueling stations are privately owned or are located at government sites closed to the public (e.g., military bases). Of the public CNG refueling stations, many require a keycard or other prior arrangement with the station operator.

[9] DOE, Alternative Fuels and Advanced Vehicles Data Center (AFDC), *Alternative Fueling Station Locator.* http://www.eere.energy.gov/afdc/fuels/stations_locator.html. Accessed September 16, 2008.

[10] "The natural-gas alternative: The pros & cons of buying a CNG-powered Honda Civic," *Consumer Reports,* April 2008. Stephe Yborra, NGVAmerica, *Frequently Asked Questions About Converting Vehicles to Operate on Natural Gas,* Washington, DC. Accessed October 15, 2008.
http://www.ngvc.org/pdfs/FAQs_Converting_to_NGVs.pdf. One potential impediment to this is the fact that in April 2009, FuelMaker declared bankruptcy in Canadian court. In May, Fuel Systems Solutions, Inc., purchased some of FuelMaker's assets, including the Phill brand. Reuters, "Key Developments: Fuel Systems Solutions Inc (FSYS.O)," May 28, 2009, http://www.reuters.com/finance/stocks/keyDevelopments?symbol=FSYS.O&pn=1.

[11] EPAct 2005. P.L. 109-58, Sec. 1342.

[12] Savings based on the following assumptions: 15,000 annual miles traveled (both vehicles); 29 miles per gallon (mpg) fuel economy for gasoline vehicle; 28 mpg equivalent for natural gas vehicle; $2.66 national retail average for regular gasoline; $11.25 per 1,000 cubic feet of residential natural gas; 121.5 cubic feet of natural gas per gasoline gallon equivalent. Therefore, current residential natural gas prices are roughly $1.37 per equivalent gallon. Fuel economy estimates from DOE, *Fueleconomy.gov.* Fuel price estimates are from EIA.

[13] However, it should be noted that a reduction in domestic consumption will likely not lead to a one-to-one reduction in imports, since reducing domestic consumption is also likely to reduce domestic petroleum production.

[14] DOE, AFDC, *Natural Gas Benefits.* http://www.eere.energy.gov/afdc/fuels/natural_gas_benefits.html. Accessed September 16, 2008.

[15] It should be noted that high natural gas prices may be needed to sustain some of this output. Otherwise, the United States may need to import natural gas to meet growing demand.

[16] The CNG Honda Civic is rated at 113 horsepower (hp), while the gasoline Civic EX is rated at 140 hp (both have 1.8 liter engines). Cars.com vehicle comparison. http://www.cars.com/go/index.jsp. Accessed September 16, 2008.

[17] 170 miles for the Civic GX vs. 345 miles for the gasoline Civic. DOE, *Fueleconomy.gov Website.* Accessed September 16, 2008.

[18] All current natural gas vehicles are modified versions of conventional gasoline vehicles. Presumably, if there were enough consumer demand, a natural gas vehicle designed from the ground up could address the problem of cargo capacity.

[19] NGV Conversion, Inc., *Frequently Asked Questions.* Accessed October 10, 2008. http://ngvus.com/p/index.html.

[20] 42 U.S.C. 7522(a)(3)(A).

[21] EPA, Office of Enforcement and General Counsel, *Mobile Source Enforcement Memorandum 1A,* June 25, 1974.

[22] EPA, Office of Enforcement and Compliance Assistance, *Addendum to Mobile Source Enforcement Memorandum 1A,* September 4, 1997.

[23] To make a conversion kit that would work for all vehicles, a manufacturer would need to certify the emissions of the conversion on every engine family for all model years—an very expensive proposition.

[24] Stephe Yborra, op. cit.

[25] NGVAmerica, *Guide to Available Natural Gas Vehicles and Engines,* Updated November 11, 2009. http://www.ngvc.org/pdfs/marketplace/MP.Analyses.NGVs-a.pdf.

[26] Stephe Yborra, op. cit.

[27] Ibid.

CHAPTER SOURCES

Chapter 1 - This is an edited, reformatted and augmented version of a Congressional Research Service publication, R41543, dated December 22, 2010.

Chapter 2 - This is an edited, reformatted and augmented version of a Congressional Research Service publication, R41027, dated January 19, 2010.

Chapter 3 - This is an edited, reformatted and augmented version of a U.S. Government Accountability Office publication, GAO-11-34, dated October, 2010.

Chapter 4 - This is an edited, reformatted and augmented version of a Congressional Research Service publication, R40963, dated November 30, 2009.

Chapter 5 - This is an edited, reformatted and augmented version of a Congressional Research Service publication, RL32205, dated December 14, 2009.

Chapter 6 - This is an edited, reformatted and augmented version of a Congressional Research Service publication, RS22971, dated February 3, 2010.

INDEX

A

access, 19, 30
accounting, 2, 4, 110
adverse effects, 3
agencies, 50, 51, 52, 65, 69, 72, 74, 75, 81, 85, 89, 90, 95, 96, 98, 108, 113, 117, 118, 119, 120, 121, 122, 123, 125, 126, 127, 128, 130, 136, 141
air emissions, 23, 72, 82, 83
air pollutants, 73, 74, 82
air quality, 48, 50, 52, 54, 61, 65, 67, 68, 69, 70, 71, 73, 74, 81
airports, 100
Al Qaeda, 100
Alaska, v, viii, 85, 86, 87, 88, 89, 90, 91, 92, 93, 94, 95, 96, 97, 98, 99, 100, 101, 102, 103, 104, 105, 111, 129, 136
Algeria, 4, 7, 12, 13, 14, 109, 111, 113, 129, 132
ambient air, 45
ambient air temperature, 45
American Recovery and Reinvestment Act, 129
American Recovery and Reinvestment Act of 2009, 129
Appropriations Act, 135
Arctic National Wildlife Refuge, 88
Argentina, 11, 13
Asia, 4, 9, 91, 134
assessment, 3, 35, 66, 67, 82, 87, 114, 123
assets, 69, 126, 135, 142
atmosphere, viii, 47, 50, 52, 53, 63, 64, 83
audit, 51, 75
authorities, 88, 108, 117, 122, 124, 125
authority, 10, 51, 66, 69, 107, 108, 109, 117, 118, 119, 120, 121, 122, 123, 124, 125, 130, 133
automation, 60
automobiles, 139
awareness, 63, 129

B

banking, 99
bankruptcy, 142
barriers, 60, 63, 129
base, 1, 7, 15, 20, 36, 40, 115
benefits, 29, 30, 38, 62, 63, 68, 69, 86, 100, 101, 129, 139, 142
biomass, 39, 41, 43, 129
boilers, 88
Bolivia, 7
Bureau of Land Management, 49, 54, 72, 103
burn, 26, 42, 112
buyer, 9
buyers, 7, 8, 9, 36

C

carbon, vii, 1, 10, 11, 17, 18, 19, 23, 24, 38, 50, 51, 59, 62, 64, 75, 81, 82, 83, 86, 97, 101, 110, 141
carbon dioxide, vii, 1, 10, 17, 23, 50, 51, 59, 64, 75, 81, 86, 97, 101, 110
carbon emissions, 1, 18, 86, 101
carcinogen, 52
cargoes, 113, 116, 127
cartel, 7
case studies, 75
case study, 126
cash, 51
catastrophic failure, 124
CEC, 134
certificate, 96, 102, 119, 124
certification, 119, 121, 124, 134
challenges, 68, 97, 121, 129
chemical, 82, 101, 116, 126, 132
chemical reactions, 82
chemicals, 116
children, 116

China, 3, 4, 10, 11, 12, 13, 91
chlorine, 116
circulation, 53
cities, 116, 124
City, 29, 91, 114, 134
Clean Air Act, viii, 52, 82, 122, 132, 137, 139
clean energy, 11
cleaning, 53
clients, 83
climate, 11, 18, 82, 87, 140, 141
climate change, 11, 18, 82, 141
CO2, 18, 19, 23, 25, 26, 30, 32, 43, 45, 48, 50, 52, 62, 81, 83
coal, vii, 1, 11, 15, 17, 18, 19, 20, 21, 22, 23, 24, 25, 26, 28, 29, 30, 31, 32, 33, 34, 35, 36, 37, 38, 39, 41, 42, 43, 44, 45, 46, 64, 97, 110, 139, 141
Coast Guard, 107, 108, 119, 120, 122, 126, 127, 128, 130, 131, 132, 135
coastal region, 108
collisions, 113
combustion, vii, 1, 19, 21, 23, 42, 43, 45, 46
commerce, 95, 104, 118
commercial, 3, 8, 15, 35, 38, 43, 134
communication, 133, 135
communities, 95, 107, 108, 121, 123, 129
community, 36, 52, 107, 121, 127, 128, 130
compensation, 69
competition, 90
competitive markets, 8, 23
compliance, 41, 90, 119, 124, 127, 139, 140
composition, 62
compounds, 52, 82
compression, 91
computer, 18, 19, 30, 38, 113, 115, 128
computer simulations, 128
Conference Report, 15
configuration, 32, 43, 98
conflict, 120
Congressional Budget Office, 99, 105
consensus, 112
conservation, 89, 99, 108, 125, 129, 130
Consolidated Appropriations Act, 99, 135
constituents, 129
construction, 4, 22, 23, 29, 30, 37, 43, 46, 83, 86, 87, 88, 89, 90, 92, 93, 94, 95, 97, 98, 99, 101, 104, 113, 117, 118, 119, 121, 125, 129, 130, 131, 133
consulting, 83
consumers, 3, 4, 8, 28, 29, 30, 36, 137, 138, 139
consumption, 2, 3, 7, 35, 37, 38, 45, 82, 87, 110, 137, 139, 142
Continental, 10, 131

control measures, 71
controversial, 98, 99, 108, 122
controversies, 100
cooking, 109
cooperation, 96
coordination, 90, 108, 109
corrosion, 99
cost, 17, 19, 21, 23, 30, 32, 35, 38, 41, 43, 46, 61, 63, 65, 73, 85, 88, 90, 91, 93, 94, 96, 97, 98, 99, 118, 127, 129, 138, 139, 140
cost effectiveness, 93
cost saving, 138, 139
counsel, 96
counterterrorism, 126
Court of Appeals, 118, 125
covering, 28, 29, 51, 81
credit market, 99
critical infrastructure, 100
crude oil, 8, 10, 100, 116
customers, 39
cyber-attack, 112
cycles, 43, 45
cycling, 40, 41, 46

D

data collection, 72
database, 22, 25, 27, 35, 37, 44, 45, 54, 67, 72, 89, 93, 131
decision makers, 115
defendants, 116
dehydration, 48, 52, 81
Delta, 87, 98, 105
demographic characteristics, 117
denial, 124
Denmark, 94
Department of Commerce, 124
Department of Energy, 44, 51, 73, 75, 82, 99, 103, 105, 114, 128, 133, 142
Department of Homeland Security (DHS), 10, 126, 130, 135
Department of the Interior, viii, 47, 49, 72, 76, 103
Department of Transportation (DOT), 10, 103, 107, 114, 117, 119, 120, 121, 123, 130, 131, 132, 133
deposits, 94, 100, 110
destruction, 36
detection, 81, 82
direct cost, 38
direct costs, 38
direct measure, 52, 59
disclosure, 129
dispersion, 117

displacement, viii, 17, 18, 19, 25, 26, 28, 29, 30, 32, 33, 34, 35, 36, 37, 38, 44, 45
disposition, 3, 69
distress, 113
distribution, iv, 39, 52, 71, 82, 93, 110
District of Columbia, 121
domestic markets, 110
draft, 70, 135
drawing, 86
drinking water, 15, 46

E

earnings, 94
economic consequences, 82
economic development, 86, 93, 101
economic downturn, 2, 7, 94
economic growth, 110
economic incentives, 62
economics, 46, 62, 71, 85, 89, 93, 94, 95, 101
education, 73
Egypt, 2, 7, 11, 12, 13, 14, 109
electricity, vii, 17, 19, 21, 23, 24, 25, 28, 29, 30, 35, 37, 39, 40, 41, 42, 43, 44, 45, 53, 110, 112, 121
emergency, 118, 131, 133
emergency response, 118, 133
emission, 18, 45, 53, 58, 59, 60, 61, 62, 66, 68, 70, 71, 74, 75, 82, 83
employees, 95, 129
energy, vii, viii, 1, 2, 3, 10, 11, 18, 21, 22, 39, 43, 44, 45, 49, 82, 85, 86, 87, 88, 89, 90, 95, 97, 100, 101, 104, 107, 108, 109, 110, 122, 123, 124, 126, 129, 130, 134, 139, 141, 142
energy consumption, 2
energy density, 139
Energy Policy Act of 2005, 90, 107, 118, 121, 122, 124, 125, 126, 127, 129, 134
energy prices, 86, 87, 97, 101
energy supply, 88, 100, 122
enforcement, viii, 118, 121, 137, 140, 141
engineering, 19, 38, 45, 62, 63, 85, 91, 96, 118, 119, 128
England, 133
environment, 36, 63, 65, 86, 94, 97, 101, 117, 127, 129, 131
environmental aspects, 67, 116
environmental effects, 99
environmental impact, 23, 38, 52, 87, 97, 98, 99, 100, 119, 122, 136, 137
environmental protection, 131
Environmental Protection Agency (EPA), 3, 11, 15, 34, 45, 47, 48, 49, 50, 51, 52, 53, 54, 55, 56, 58, 59, 60, 61, 62, 63, 64, 65, 68, 69, 70,
71, 72, 73, 74, 75, 81, 82, 83, 116, 119, 132, 133, 137, 139, 140, 141, 142
Equatorial Guinea, 14, 109
equipment, 8, 48, 52, 53, 55, 58, 61, 63, 64, 66, 68, 81, 82, 83, 95, 110, 112, 138, 140
erosion, 118
Europe, 4, 7, 11
evidence, 51, 75, 76, 118
exclusion, 59, 113, 117, 123, 131
Executive Order, 89, 133
exercise, 108, 109, 129
expenditures, 118, 127
expertise, 63, 114
explosives, 100, 112
exporter, 2, 4
exports, 4, 10, 14, 89, 109
exposure, 116
extraction, 61, 131

F

families, 116, 140
fear, 107, 113, 121, 130
fears, 111
federal agency, 99
federal authorities, 100, 126
federal government, 49, 51, 52, 63, 69, 82, 99, 101, 108, 120, 122, 125, 133
federal law, 108, 119, 124
federal permitting, 124
Federal Register, 105, 134
federal regulations, 108, 118
Federal Reserve Board, 131
financial, 62, 90, 98, 99, 116, 136
financial markets, 62
financial support, 98
fire detection, 113
fires, 112, 114, 126
First Amendment, 129
fiscal year 2009, 49
flame, 45
flexibility, 4, 17, 38, 65, 140
fluctuations, 40
force, 89, 103
Ford, 140, 142
forecasting, 36
formation, 61
formula, 8
fractures, 15
France, 11, 110, 132
freezing, 62, 112
fuel prices, 46, 138
funding, 126, 127
funds, 90, 92

G

gasification, 43
General Accounting Office (GAO), viii, 47, 48, 51, 54, 56, 57, 58, 61, 70, 71, 72, 75, 81, 82, 83, 115, 123, 127, 128, 132, 134, 135, 143
General Motors, 140, 142
geology, 62
Georgia, 110, 121
Germany, 11
global climate change, 50
global warming, 52, 64
glycol, 53, 55, 66, 70, 71, 82
government funds, 49
governor, 92, 102, 122, 131
grants, 107, 118, 141
greenhouse, vii, viii, 17, 18, 47, 48, 50, 52, 59, 60, 63, 64, 69, 70, 71, 72, 73, 75, 81, 82, 105, 139, 141
greenhouse gases, 50, 52, 63, 64, 69, 71, 73, 81
grids, 29, 112
grounding, 113
growth, 10, 11, 20, 23, 36, 86, 94, 97, 108, 110, 129, 130, 139
guidance, 47, 48, 50, 55, 58, 59, 65, 67, 69, 70, 71, 75, 81, 82, 83, 90, 120, 126, 140
guidelines, 51, 117
Guinea, 7
Gulf Coast, 3
Gulf of Mexico, 36, 50, 54, 58, 59, 73, 82, 101, 110

H

harbors, 116
Hawaii, 91
hazards, 101, 107, 108, 109, 111, 112, 113, 114, 115, 116, 121, 128, 129, 130
haze, 52
health, 95, 120
history, 87, 93, 97, 100, 134
homeland security, 116
Homeland Security Act, 10
homes, 55, 109
House, 49, 104, 105, 108, 119, 131, 132, 135, 141
House of Representatives, 49
hub, 8, 91, 92
human, 131
human welfare, 131
Hunter, 132
Hurricane Katrina, 82
hydrocarbons, 52, 81

I

ignition source, 112
imports, 2, 4, 7, 10, 11, 35, 36, 97, 107, 108, 109, 110, 127, 130, 139, 141, 142
improvements, 3, 35, 64, 111, 129
India, 11
individuals, 114
Indonesia, 12, 13, 14, 109
industries, 44, 104, 111, 131
industry, viii, 3, 8, 9, 10, 11, 15, 18, 23, 37, 47, 48, 50, 51, 52, 53, 59, 61, 62, 63, 65, 66, 68, 71, 73, 74, 75, 81, 82, 83, 100, 108, 109, 110, 113, 116, 120, 125, 130
inertia, 63
information exchange, 68
infrastructure, 18, 38, 62, 86, 94, 95, 98, 101, 108, 109, 112, 116, 119, 122, 123, 126, 127, 129, 130, 138, 141
injure, 112, 113, 126
inspections, 51, 66, 67, 83, 118, 119
inspectors, 66, 68, 83
integrity, 124, 129
intelligence, 127
interest rates, 97
investment, 24, 62, 63, 95, 97
investments, 62, 63, 64
investors, 93
Iran, 3, 7, 11, 12, 13
Iraq, 12
irrigation, 46
issues, vii, 17, 19, 26, 38, 43, 45, 82, 87, 97, 102, 104, 105, 107, 108, 119, 120, 122, 128, 130, 131
Italy, 11, 124

J

Japan, 11, 14, 89, 102, 109, 110, 124
job creation, 95
jurisdiction, 10, 117, 119, 120, 121

K

Kazakhstan, 4, 12
Korea, 124
Kuwait, 12

L

labor force, 95
law enforcement, 117
laws, 51, 52, 120, 121
lead, 29, 30, 37, 64, 99, 114, 118, 124, 142

leaks, 49, 52, 67, 83
legislation, 11, 62, 66, 69, 87, 90, 108, 109, 122, 129
legislative proposals, 96, 122
legs, 88
lifetime, 95
light, 52, 93, 138, 140
limestone, 15
liquefied natural gas, 10, 35, 86, 108, 109, 117, 128, 136, 138
liquids, 10, 51, 53, 81, 92, 103
loan guarantees, 85, 99
lobbying, 128
local government, 94, 118, 128, 130
Louisiana, 75, 110
lower prices, 93
LPG, 140

M

magnitude, 75
majority, 3, 20, 30, 58, 139
Malaysia, 12, 13, 14, 109
man, 100
management, 51, 63, 75, 131
manufacturing, 15, 35
marketability, 139
marketplace, 142
Maryland, 110, 113, 125
materials, 83, 98
matter, iv, vii, 1, 103
measurement, 50, 59, 83
measurements, 72
media, 116
medical, 117
mercury, 43
meter, 59, 68
methodology, 51
Mexico, 11, 13, 14, 57, 73, 74, 75, 91, 116
Middle East, 10
military, 100, 142
mission, 127
missions, 127
misunderstanding, 100
mixing, 112
models, 30, 59, 113, 114, 115, 123, 128, 142
modifications, 85, 118, 139
moisture, 81
momentum, 85, 101
Montana, 29
motivation, 87

N

national policy, 86, 97, 110
National Research Council, 105
national security, 124
natural disaster, 101
natural disasters, 101
Netherlands, 11, 13, 132
New England, 134
Nigeria, 7, 12, 14, 109
nitrogen, vii, 1, 43, 74, 81
nitrous oxide, 81
North America, 93, 94, 97, 98, 100, 104, 105, 108, 109, 111, 129, 130, 131
Norway, 4, 12, 13, 14, 109

O

Obama, 86, 102
obstacles, 85, 101
officials, viii, 47, 51, 53, 58, 59, 61, 62, 63, 64, 66, 67, 68, 69, 73, 75, 82, 94, 96, 107, 118, 121, 123, 127, 128
oil, vii, viii, 1, 9, 10, 15, 23, 43, 45, 46, 47, 48, 49, 50, 51, 52, 53, 54, 55, 59, 61, 62, 64, 65, 67, 68, 69, 70, 72, 73, 74, 75, 81, 82, 83, 95, 99, 100, 101, 102, 112, 113, 114, 116, 126, 130, 131, 137, 141
oil production, 62, 73, 95, 100, 101
oil revenues, 95
oil sands, 100
oil spill, 113, 116
Oklahoma, 3, 28
operating costs, 40
operations, 45, 51, 63, 66, 74, 81, 82, 83, 95, 100, 113, 118
opportunities, iv, 60, 62, 64, 66, 67, 68, 69, 70, 71, 75, 83, 89, 141
optimism, 35, 36
organic compounds, 52, 56
Outer Continental Shelf Lands Act, 10, 51
outreach, 73
oversight, 45, 48, 51, 65, 69, 75, 86, 101, 108, 109, 119, 123, 126, 127, 131
ownership, 130
ozone, 52, 61, 83

P

Pacific, 29, 88, 89, 91, 102, 134
participants, 65, 68, 82, 121
payback period, 138, 139
penalties, 116
permit, 23, 52, 89, 99, 121, 124, 125

perpetrators, 100
petroleum, 10, 21, 43, 116, 137, 139, 140, 141, 142
Petroleum, 15, 50, 74, 81, 91, 102, 104
Philadelphia, 105
physical phenomena, 113
physical properties, 124
plants, vii, 17, 18, 19, 20, 21, 22, 23, 24, 25, 26, 28, 29, 30, 31, 32, 33, 34, 35, 37, 38, 39, 40, 41, 42, 43, 44, 45, 46, 64, 113, 116, 121
platform, 66, 73
police, 100
policy, vii, 1, 17, 18, 19, 36, 38, 39, 68, 74, 82, 87, 89, 93, 96, 97, 99, 100, 101, 103, 105, 108, 109, 118, 122, 124, 126, 127, 128, 129, 130, 137, 141
policy issues, 87, 96, 108
policy makers, 93, 96, 97, 99, 122, 124, 126, 127, 128, 129, 130
policy making, 82
pollutants, 124, 139, 141
pollution, 67
population, 29, 52, 117, 126, 132
portfolio, 49
power generation, 3, 11, 23, 37, 46, 86, 97, 109
power lines, 28, 29
power plants, 17, 18, 19, 20, 21, 22, 23, 24, 26, 29, 30, 33, 36, 38, 39, 42, 64, 88, 97, 100, 110, 130
preparation, iv, 99, 119
present value, 93, 94
president, 87
President, viii, 43, 85, 86, 87, 89, 97, 98, 103, 129, 135
President Obama, viii, 85, 86, 97, 129
prevention, 117, 120
price ceiling, 10
price manipulation, 7
probability, 35, 115, 126, 127, 135
producers, 4, 9, 22, 23, 48, 50, 52, 59, 68, 69, 72, 74, 81, 90, 91, 92, 93, 94, 96, 140
production costs, 97
profit, 66, 100
progress reports, 90
project, viii, 67, 85, 86, 87, 88, 90, 91, 92, 93, 94, 95, 96, 97, 98, 99, 100, 101, 102, 124, 126, 130, 133, 134
project sponsors, 92
propane, 116
proposition, 142
protection, 131
prototype, 43
public concern, 107, 108, 121, 130

public concerns, 107, 108, 130
public education, 10
public interest, 133
public safety, 108, 109, 115, 116, 119, 120, 122, 123, 128, 130
public sector, 127
Puerto Rico, 110, 121

Q

quality standards, 67, 124
questioning, 127

R

radar, 113
radiation, 112, 113
Radiation, 70, 71, 132
radioactive waste, 123
radius, 30, 31, 33, 45
ramp, 40, 41
recession, 30, 35, 129
recommendations, iv, 47, 50, 53, 59, 70, 71
recovery, 3, 15, 42, 53, 61, 62, 63, 66, 67, 68, 75, 83, 90
recreation, 46
recreational, 125
Reform, 49
regulations, 11, 45, 47, 48, 50, 51, 52, 58, 61, 65, 66, 69, 70, 71, 74, 81, 82, 107, 108, 109, 116, 117, 118, 119, 120, 121, 122, 123, 130, 131, 139
regulatory agencies, 125
regulatory changes, 61
regulatory requirements, 98, 133
reimburse, 118
reliability, 29, 32, 33, 46, 50, 119
relief, 124
renewable energy, 129, 130
repair, 64, 75
requirements, 32, 33, 59, 61, 67, 71, 90, 91, 93, 97, 99, 107, 108, 117, 118, 120, 121, 122, 123, 127, 129, 130, 131, 133, 140
reserves, 2, 3, 4, 15, 18, 86, 87, 91, 92
resistance, 44
resolution, 87
resource allocation, 127
resources, vii, 1, 2, 3, 8, 11, 15, 18, 35, 36, 42, 51, 62, 65, 69, 86, 87, 90, 98, 99, 107, 109, 122, 127, 128
response, 52, 59, 71, 88, 128, 137
restrictions, 30
retail, 138, 142
revenue, 49, 50, 52, 95, 103

risk, 62, 94, 99, 117, 118, 119
risks, 21, 23, 90, 97, 99, 115, 122, 126
routes, 87
royalty, viii, 47, 48, 50, 51, 52, 60, 63, 64, 69, 72, 75, 82, 83, 95
rules, 99, 119, 125
runoff, 43
rural areas, 52
Russia, 3, 4, 7, 8, 11, 12, 13, 14

S

sabotage, 123
safety, 10, 21, 49, 107, 108, 109, 111, 112, 113, 114, 115, 116, 117, 118, 119, 120, 121, 122, 123, 124, 127, 128, 129, 130, 131
safety zones, 108, 131
Saudi Arabia, 11, 12, 13
savings, 62, 63, 138
school, 15
scope, 19, 30, 38, 51, 71, 100, 113, 114
Secretary of Commerce, 124
Securities Exchange Act, 133
security, 10, 23, 86, 87, 100, 101, 105, 107, 108, 109, 116, 118, 119, 120, 122, 126, 127, 130, 131
security forces, 128
self-assessment, 36
seller, 9
sellers, 7, 8
Senate, 18, 36, 43, 45, 86, 99, 103, 134, 135, 141
settlements, 116
shape, 40
signals, 36
signs, 83
sludge, 43
software, 59, 132
solid waste, 43
solution, 18, 30
South Korea, 14, 110
Spain, 110, 124
spending, 18, 95
stability, 46
stabilization, 45
stakeholders, 85, 93, 95, 99, 100, 101, 114, 130
state, 3, 10, 15, 23, 51, 52, 53, 61, 64, 69, 74, 75, 81, 82, 86, 91, 92, 94, 95, 96, 103, 107, 108, 109, 116, 118, 120, 121, 122, 123, 124, 125, 128, 130, 131, 134, 137
state oversight, 10
states, viii, 17, 26, 28, 29, 38, 52, 53, 55, 61, 67, 72, 74, 85, 86, 87, 89, 90, 92, 93, 94, 95, 96, 97, 98, 99, 101, 107, 108, 109, 115, 118, 119, 121, 122, 124, 125, 126, 127, 128, 129, 132, 135, 139
statistics, 44
statutes, 10, 88, 117, 119, 125, 126
steel, 98
storage, 10, 17, 26, 37, 38, 40, 42, 52, 53, 55, 56, 57, 59, 61, 62, 63, 67, 73, 75, 80, 81, 82, 83, 110, 113, 116, 121, 123, 124, 127, 130, 139
stress, 59
structural changes, 94
structure, 7, 115, 130
subsidy, 99
sulfur, vii, 1, 43, 81
sulfur dioxide, vii, 1, 43
suppliers, 109, 140
supply chain, 126
Supreme Court, 88, 125, 129, 133
surplus, 25, 28, 30, 33, 44
Sweden, 94

T

Taiwan, 110
tanks, 52, 53, 55, 56, 57, 59, 61, 63, 67, 75, 80, 81, 82, 83, 110, 113, 116, 124, 139
tar, 100
target, 94, 112
tariff, 90
tax incentive, 138, 141
teams, 62
technical comments, 70
techniques, 3, 52, 59
technologies, viii, 18, 46, 47, 48, 50, 51, 53, 60, 61, 62, 63, 64, 65, 67, 68, 69, 70, 71, 72, 73, 74, 75, 82, 83, 87, 123
technology, 8, 15, 18, 19, 20, 21, 22, 23, 29, 42, 43, 44, 46, 48, 49, 52, 60, 61, 63, 67, 69, 70, 75, 81, 123
temperature, 52, 112
tension, 125
terminals, viii, 15, 107, 108, 109, 110, 111, 113, 116, 117, 118, 120, 121, 122, 123, 124, 125, 126, 127, 128, 129, 130, 131
territorial, 95
territory, 29, 87
terrorism, 97, 107, 108, 112, 122, 127
terrorists, 100, 107, 112, 126, 127, 129
testing, 128, 136, 140
Thailand, 11
threats, 101, 126, 127, 130, 131
time frame, 83, 97
Title I, 105, 135
Title II, 135
total energy, 42

total product, 72
total revenue, 93
tracks, 15, 95
trade, 4, 11
training, 116
transmission, 10, 17, 18, 19, 25, 26, 28, 29, 30, 32, 38, 39, 44, 46, 89, 109, 110, 130
transparency, 129
transport, 4, 88, 89, 91, 92, 94, 98, 99, 100
transportation, viii, 4, 10, 15, 26, 32, 35, 37, 81, 86, 87, 88, 94, 102, 104, 108, 110, 116, 118, 121, 128, 130, 137, 141
Transportation Security Administration, 10
treatment, 85, 90, 91, 92, 96
Trinidad, 2, 7, 13, 14, 109
Turkmenistan, 4, 12

U

U.S. economy, vii, 1, 94
U.S. Geological Survey, 87, 102, 105
Ukraine, 11
unacceptable risk, 108, 109, 115
United, v, vii, viii, 1, 2, 8, 10, 11, 12, 13, 14, 20, 28, 43, 44, 46, 47, 52, 74, 81, 82, 87, 88, 91, 97, 98, 100, 102, 107, 108, 109, 110, 120, 125, 126, 133, 134, 136, 138, 139, 142
United Kingdom, 8, 11, 13
United Nations, 82
United States, v, vii, viii, 1, 2, 8, 10, 11, 12, 13, 14, 20, 28, 43, 44, 46, 47, 52, 74, 81, 87, 88, 91, 97, 98, 100, 102, 107, 108, 109, 110, 120, 125, 126, 133, 134, 136, 138, 139, 142
up-front costs, 137, 138
urban, 116, 124
Uzbekistan, 4, 11, 13

V

validation, 115, 128
valve, 75, 82
vandalism, 100
vapor, 53, 61, 62, 63, 66, 68, 75, 112, 113, 114, 117

variable costs, 41, 42, 46
variables, 19
vehicles, viii, 64, 97, 110, 137, 138, 139, 140, 141, 142
vein, 63
Venezuela, 7, 12
vessels, 83, 120
veto, 122
Vice President, 43, 103, 104
volatile organic compounds, 52, 71, 82
volatility, viii, 86, 89, 97, 107, 108
vulnerability, 97, 101, 123

W

Wales, 132
Washington, 75, 81, 83, 105, 116, 131, 132, 133, 134, 135, 142
waste, 22, 42, 44, 51, 65, 66, 69, 70, 83, 121
waste disposal, 121
waste heat, 22, 42
water, 3, 15, 46, 52, 81, 112, 114, 115, 116, 121, 124, 125, 128, 131, 134
water quality, 124, 125, 134
water quality standards, 124
water rights, 134
water supplies, 3
water vapor, 81
waterways, 120
weakness, 94
weapons, 100, 112
wells, 15, 52, 60, 62, 63, 67, 73, 83, 129
wholesale, 8
wilderness, 86, 101
wildlife, 99
Wisconsin, 102
wood, 20
workers, 95, 111, 113
worldwide, 113, 134

Y

yield, 59, 85, 96